Lecture Notes in Economics and Mathematical Systems

623

Christian Ullrich

Forecasting and Hedging in the Foreign Exchange Markets

 Springer

Dr. Christian Ullrich
BMW AG
Petuelring 130
80788 München
Germany
christian.ullrich@bmw.de

HG
3851
. U45
2009

ISSN 0075-8442
ISBN 978-3-642-00494-0   e-ISBN 978-3-642-00495-7
DOI 10.1007/978-3-642-00495-7
Springer Dordrecht Heidelberg London New York

Library of Congress Control Number: 2009926055

Cover design: SPi Publisher Services

Printed on acid-free paper

Springer is part of Springer Science+Business Media (www.springer.com)

For Robin.

Foreword

Historical and recent developments at international financial markets show that it is easy to loose money, while it is difficult to predict future developments and optimize decision-making towards maximizing returns and minimizing risk. One of the reasons of our inability to make reliable predictions and to make optimal decisions is the growing complexity of the global economy. This is especially true for the foreign exchange market (FX market) which is considered as one of the largest and most liquid financial markets. Its grade of efficiency and its complexity is one of the starting points of this volume.

From the high complexity of the FX market, Christian Ullrich deduces the necessity to use tools from machine learning and artificial intelligence, e.g., support vector machines, and to combine such methods with sophisticated financial modeling techniques. The suitability of this combination of ideas is demonstrated by an empirical study and by simulation. I am pleased to introduce this book to its audience, hoping that it will provide the reader with interesting ideas to support the understanding of FX markets and to help to improve risk management in difficult times.

Moreover, I hope that its publication will stimulate further research to contribute to the solution of the many open questions in this area.

Karlsruhe, January 2009   *Detlef G. Seese*

Preface

The growing complexity of many real world problems is one of the biggest challenges of our time. The area of international finance is one prominent example where decision making is often fraud to mistakes, and tasks such as forecasting, trading and hedging exchange rates seem to be too difficult to expect correct or at least adequate decisions. Could it be that it is too complex for decision-makers to arrive at an optimal solution from a computational complexity perspective?

As a first task we address the problem of forecasting daily exchange rate directions. We challenge the widely believed idea of efficient markets as offered by modern finance theory and represent the problem formally as a classification task according to the binary classification problem (BCP), which is known to be NP-complete. The BCP is addressed with support vector machines (SVM), a state-of-the-art supervised learning system that has emerged from the fields of computational geometry and computational statistics and has proven to successfully exploit patterns in many nonfinancial applications. Six SVM models with varying standard kernels, along with one exotic p-Gaussian SVM are compared to investigate the separability of Granger-caused input data in high-dimensional feature space. To ascertain their potential value as out-of-sample forecasting and quantitative trading tool, all SVM models are benchmarked against traditional forecasting techniques. We find that hyperbolic SVMs perform well in terms of forecasting accuracy and trading performance via a simulated strategy. Moreover, p-Gaussian SVMs perform reasonably well in predicting EUR/GBP and EUR/USD directions.

Second, we address the fact that as business has become global, firms must protect themselves from the potential losses arising from fluctuating exchange rates. Given a stochastic representation of the exchange rate, we investigate the problem of optimally hedging foreign exchange exposure with a combination of linear and nonlinear financial contracts such that the expected utility at the planning horizon is maximized. A weighted mean–variance–skewness utility maximization framework with linear constraints is embedded in a single-period stochastic combinatorial optimization problem (SCOP) formulation in order to find optimal combinations for spot, forward, and European style straddle option contracts. The problem is shown to be NP-complete. In order to derive near-optimal decisions within a reasonble amount of time, a simulation/optimization procedure is suggested. Exchange rate behavior is modeled via a nonlinear smooth transition PPP reversion model. Optimization is carried out with a variant of the scatter search metaheuristic. Dynamic backtesting with real-world financial data is conducted to validate the model and to show its applicability in a practical context. We find in our experiments that scatter search is a search method that is both aggressive and robust. The simulation/optimization approach adds value in terms of reducing risk and enhancing income in comparison to several passive strategies.

The dissertation was supervised by Prof. Dr. D. Seese, Institute AIFB, University of Karlsruhe, and has been written in English language.

München, January 2009   *Christian Ullrich*

Acknowledgements

I am most grateful to the kind support, the trenchant critiques, and the probing questions during the completion of this thesis of my thesis supervisor Prof. Dr. Detlef Seese. Sincere thanks to my reviewers Prof. Dr. Marliese Uhrig-Homburg, and Prof. Dr. Svetlozar Rachev for their input and interest in this research. Their detailed comments were of great importance in refining the publications and the draft version of the thesis to its final form. I am very grateful for the fruitful discussions with my fellow doctoral students Thomas Stümpert, Andreas Mitschele, Tobias Dietrich, and Amir Safari from the Complexity Management Research Group at the AIFB Institute, as well as Dr. Stephan Chalup from University of Newcastle's Robotic Laboratory.

Since this dissertation was mainly written during my time as a doctoral candidate in BMW Group's financial risk management department, I am not less than indebted to my supervisors Norbert Mayer for his ingenuousness, constant curiosity, and interest in innovation and ongoing improvement; Joachim Herr for his patience, dedicated guidance, quick grasp, and enviable promotional talents; Ron Tonuzi for providing a stimulating and fun environment in which to learn and grow. I would also like to thank my colleagues Karl Ertl and Peter Studtrucker for sharing valuable insights and experiences in practical currency trading and soccer betting. This dissertation also drew upon the knowledge, insights and experience of numerous BMW internals as well as external business partners from the banking and consulting industries. Although too numerous to name, they are appreciated for their individual contributions and inspiration.

Thank you also to the musicianship and different sides of my band who enabled me to have a wonderful time along the way. I am particularly indebted to one guy whose creative endeavors, coolness, and wisdom allowed me to maintain my (in)sanity throughout the writing of this thesis: Bob Dylan. I want to thank my parents, who have always supported my educational pursuits and whose foresight and values paved the way with unconditional love and support. Finally, I want to thank my girlfriend Robin for her love and patience.

Contents

Acronyms

ACF	Autocorrelation function
ADF	Augmented Dickey–Fuller
AIC	Akaike's information criterion
ANN	Artificial neural network
AR	Autoregressive
ARCH	Autoregressive conditional heteroskedasticity
AR(I)MA	Autoregressive (integrated) moving average
BCP	Binary classification problem
BDS	Brock–Dechert–Scheinkmann
CIA	Covered interest arbitrage
CIRP	Covered interest rate parity
CPI	Consumer price index
DCOP	Deterministic combinatorial optimization problem
DW	Durbin–Watson
ERM	Empirical risk minimization
ESTAR	Exponential smooth transition autoregressive
FX	Foreign exchange
GARCH	Generalized autoregressive conditional heteroskedasticity
IID	Independent identically distributed
IRP	Interest rate parity
JB	Jarque–Bera
LM	Lagrange-multiplier
KKT	Karush–Kuhn–Tucker
KPSS	Kwiatkowski–Phillips–Schmidt–Shin
KU	Kurtosis
LB	Ljung-Box
LP	Linear programing
LSTAR	Logistic smooth transition autoregressive
MA	Moving average
MCS	Monte Carlo simulation
MEH	Market efficiency hypothesis

NID	Normally identically distributed
OLS	Ordinary least squares
PACF	Partial autocorrelation function
PCA	Principal component analysis
PCOP	Probabilistic combinatorial optimization problem
PP	Phillips–Perron
PPP	Purchasing power parity
PTSP	Probabilistic traveling salesman problem
QP	Quadratic programing
RBF	Radial basis function
SCOP	Stochastic combinatorial optimization problem
SETAR	Self-exciting threshold autoregressive
SIP	Stochastic integer programing
SK	Skewness
SMO	Sequential minimization problem
SRM	Structural risk minimization
STAR	Smooth transition autoregressive
SV	Support vector
SVC	Support vector classification
SVM	Support vector machines
TAR	Threshold autoregressive
TSP	Traveling salesman problem
UIA	Uncovered interest arbitrage
UIRP	Uncovered interest rate parity
VAR	Vector autoregressive
VC	Vapnik–Chervonenkis

Part I
Introduction

Chapter 1
Motivation

The growing complexity of many real-world problems is one of the biggest challenges of our time. Complexity is also one of the factors influencing decision making in the area of international finance which has taken on great significance over the last decade. As the result of structural shifts in the world economy and in the international financial system, the foreign exchange market has been profoundly transformed, not only in size, but also in coverage, architecture, and mode of operation. Among the major developments that have occurred in the global financial environment are the following:

1. A basic change in the international monetary system from the fixed exchange rate requirements of Bretton Woods that existed until the early 1970s to the floating exchange rate system of today.
2. A tidal wave of financial deregulation throughout the world, with massive elimination of government controls and restrictions in nearly all countries, resulting in greater freedom for national and international financial transactions, and in greatly increased competition among financial institutions, both within and across national borders.
3. A fundamental move towards institutionalization and internationalization of savings and investment, with funds managers and institutions having vastly larger sums available, which they are investing and diversifying across borders and currencies in novel ways.
4. A broadening and deepening trend towards international trade liberalization, within a framework of multilateral trade agreements, such as the Tokyo and the Uruguay Rounds of the General Agreement on Tariffs and Trade, the North American Free Trade Agreement and U.S. bilateral trade initiatives with China, Japan, India and the European Union.
5. Major advances in technology, making possible instantaneous real-time transmission of vast amounts of market information worldwide, immediate and sophisticated manipulation of that information in order to identify and exploit market opportunities, and rapid and reliable execution of financial transactions.
6. Breakthroughs in the theory and practice of finance, resulting not only in the development of new financial instruments and derivative products, but also in

C. Ullrich, *Forecasting and Hedging in the Foreign Exchange Markets,* Lecture Notes in Economics and Mathematical Systems 623, DOI: 10.1007/978-3-642-00495-7_1,
© Springer-Verlag Berlin Heidelberg 2009

advances in thinking that have changed market participants' understanding of the financial system and their techniques for operating within it.

The interplay of these forces, feeding off each other in a dynamic and synergistic way, led to an environment where foreign exchange trading increased rapidly. With world-wide daily trading at approximately three trillion US Dollars [39], the Foreign Exchange market is by far the largest and most liquid market in the world, where liquidity refers to currencies' ability to be easily converted through an act of buying or selling without causing a significant movement in the price and with minimum loss of value.

While an exchange rate qualifies as a security because it is traded daily, it generates no cash flow like a bond, offers no dividend like a stock, and has neither a discounted present value nor a terminal value as in any typical asset. Currencies have been termed a *medium of exchange* and thus while traded like a security, they are not an asset in the true sense of the word. Hence, one should not expect exchange rates to have a return as they are just the *grease* to exchange products across different geographical borders. However, because there are various forces that affect the supply and demand of currencies, currency values change almost continuously, thereby generating return opportunities (positive or negative) even though theoretically they should not. Hence, if currencies are in excess demand relative to a given rate because of an influx of foreign investors, the currency will appreciate, but through changes in international trade and investment, the assets in the foreign country will be overvalued leading to either a reduction in demand or an increase in the quantity supplied leading to a subsequent depreciation.

Attempts to explain and predict exchange rate movements have largely remained unsuccessful. For instance, many researchers have tried to predict future currency levels using various structural and time series models and have come to the conclusion that currencies are extremely hard to forecast and that these models perform no better than a random walk model [287]. While there is a broad consensus about long term movements over several years being driven by fundamental economic indicators, there seem to be no reliable methods of forecasting exchange rates over medium to short time horizons such as days, weeks, or even months. In fact, currency values in the short term can be very volatile and erratic for largely inexplicable reasons. Krugman and Obstfeld [235] write:

> If exchange rates are asset prices that respond immediately to changes in expectations and interest rates, they should have properties similar to those of other asset prices, for example, stock prices. Like stock prices, exchange rates should respond strongly to "news," that is, unexpected economic and political events; and, like stock prices, they therefore should be very hard to forecast.

However, forecasting is necessary in order to give policy advice for prominent practical tasks involving trading and hedging currencies where decisions must be made without advance knowledge of their consequences.

As a first task we take a data-driven approach in order to examine whether there is any predictability in the behavior of daily exchange rate data. We challenge the widely believed idea of efficient markets as offered by modern finance theory [124, 126] by using a new supervised learning forecasting technique that has emerged from the fields of computational geometry and computational statistics. This technique is called support vector machines (SVM) and has proven to successfully exploit patterns in many nonfinancial applications. In addition, SVM handles the complexity of identifying the optimal decision function in a very interesting way. It operates in a space with possibly infinite dimensions as opposed to the well known approach of artificial neural networks (ANN), which is designed to approximate a nonlinear function in input space. The hypothesis is, that with a different quantitative approach which has the computational power to process a sufficient amount of information, behavior that was previously believed random might become predic.

Second, we address the fact that as business becomes more global, more and more nonfinancial companies are finding themselves increasingly exposed to foreign currency exposure. On average, firms judge between a quarter and a third of their revenues, costs and cashflows as being exposed to movements in exchange rates. It is therefore not a surprise that multinational firms, according to Servaes and Tufano [364], consider foreign exchange risk as the most costly type of corporate risk on a six point scale (see Table 1.1).

In the absence of any reduction in exchange rate volatility, firms must protect themselves from the potential losses arising from unexpected changes in exchange rates provided that markets are imperfect in the sense of [298] and it is not possible for the company's shareholders to hedge the risk themselves. We investigate the problem of an industrial corporation that seeks to optimally hedge its foreign exchange exposure with a combination of linear and nonlinear financial contracts

Table 1.1 "Without risk management, how costly would the following risks be to your company over the next 5 years, considering both likelihood and magnitude of loss?"

Factors	%4 or 5	N
Foreign exchange risks	53%	239
Strategic risks	47%	232
Financing risks	40%	233
Competitive risks	39%	237
Failure of company projects	36%	236
Execution risks	35%	231
Reputational risks	33%	232
Commodity price risks	32%	238
Operational risks	31%	232
Interest rate risks	31%	239
Credit risks	28%	237
Regulatory or government risks	26%	238
Loss of key personnel	26%	235
Property and casualty risks	22%	236
Litigation risks	21%	233
Natural catastrophe risks	17%	237
Employee misdeeds	13%	234
Terrorism risks	13%	235
Political risks	11%	238
Pension or healthcare shortfalls	10%	228
Weather risks	9%	235

Scale is "Not Important" (0) to "Very Important" (5)

given a stochastic representation about the underlying exchange rate. Despite a large body of research, rooted in economics, international finance, portfolio theory, statistics, and operations research having explored the optimal hedging of foreign exchange risks, newspaper headlines regularly suggest, that firms have either not found a recipe yet against adverse exchange rate fluctuations, or the applied methods and strategies do not provide sufficient protection against currency risks. Decision making in real life situations of high complexity is often fraud to mistakes, and many problems seem to be too difficult to expect correct or at least adequate decisions. Could it be that it is too complex for decision-makers to arrive at an optimal solution from a computational complexity perspective? First, firms may be exposed to different sources of uncertainty. Not only is there uncertainty related to the exchange rate of various currency pairs, but there usually exists uncertainty regarding the underlying exposures which must be estimated in order to determine the appropriate volume of a potential derivative hedge. Second, derivative markets offer a plethora of different products with different pay-off structures, and it is not obvious which product or which combination of products is most appropriate for the risk structure of a particular firm at a certain point in time. Third, decision-making under uncertainty usually involves finding a good compromise between objectives which may be contradictive and highly individual, random quantities with respect to the information at the moment, and the inclusion of personal expectations. As a consequence, there are open questions concerning the relation between the theoretical benefits of hedging, reduced volatilities, corporate performance, and the appetite for risk. For instance, it is still an open question, whether hedging can do more than just reducing potential losses and potential gains. It is therefore of both theoretical and practical interest to examine how the problem of finding a particular hedging strategy that optimally matches a firm's preferences towards risk and reward can be solved in a computationally efficient way and compares against static benchmark strategies.

Chapter 2
Analytical Outlook

2.1 Foreign Exchange Market Predictability

Research in the area of exchange rates is vast and any attempt to survey the exchange rate theory in its totality would be impossible. Consequently, the objective of Part II is more moderate. It provides a partial review of the classical concepts of exchange rate determination theory and financial theory. In the last few decades, exchange rate economics has seen a number of developments, with substantial contributions to both the theory and empirics of exchange rate determination. Important developments in econometrics and the increasing availability of high-quality data have also been responsible for stimulating the large amount of empirical work on exchange rates. We restrict ourselves to explore the fundamental parity conditions purchasing power parity (PPP) and interest rate parity (IRP) as prerequisites for the understanding of economic equilibrium along with explanations given by economists on why exchange rates in reality do often not behave according to these laws. Furthermore the notions of market efficiency are highlighted, upon which standard finance theory is built.

While our understanding of exchange rates has significantly improved, empirical studies have revealed a number of challenges and open questions in the exchange rate debate which have remained unsolved. For instance, while all of the potential explanations for deviations from equilibrium conditions appear fairly reasonable and have theoretical merit, the PPP disconnect and PPP excess volatility puzzles, along with the forward anomalie puzzle have not yet been convincingly explained and continue to puzzle the international economics and finance profession. Furthermore, it is a well-known phenomenon that the forward rate is not an unbiased predictor for the expected exchange rate as the market efficiency hypothesis postulates.

It is argued that this inexplicability has nothing to do with economic and financial theorists being not capable enough, but rather with the complex dynamics that drive exchange rates, the computational complexity of behaving according to economic theory, and the nature of complexity itself. For these reasons, selected proveable mathematical results on the difficulty of calculating market equilibrium, and the computational difficulties with market efficiency are provided.

C. Ullrich, *Forecasting and Hedging in the Foreign Exchange Markets,* Lecture Notes in Economics and Mathematical Systems 623, DOI: 10.1007/978-3-642-00495-7_2, © Springer-Verlag Berlin Heidelberg 2009

2.2 Exchange Rate Forecasting with Support Vector Machines

In Part III, we address the problem of predicting daily exchange rate directions with Support Vector Machines (SVM). The construction of machines capable of learning from experience has for a long time been the object of both philosophical and technical debate. The technical aspect of the debate has received an enormous impetus from the advent of electronic computers. They have demonstrated that machines can display a significant level of learning ability, though the boundaries of this ability are far from being clearly defined. The availability of reliable learning systems is of strategic importance. When computers are applied to solve a practical problem it is usually the case that the method of deriving the required output from a set of inputs can be described explicitly. As computers are applied to solve more complex problems, however, situations can arise in which there is no known method for computing the desired output from a set of inputs, or where that computation may be very expensive.

This motivates the use of an alternative strategy where the computer is instructed to attempt to learn the input/output functionality from examples, which is generally referred to as supervised learning, a sub-discipline of the machine learning field of research. Machine learning models are rooted in artificial intelligence which differs from economic and econometric theory since statistical inferences are made without any a priori assumptions about the data. Intelligent systems are therefore designed to automatically detect patterns, i.e., any relations, regularities or structure inherent in a given dataset, that exists despite complex, nonlinear behavior. If the patterns detected are significant, a system can be expected to make predictions about new data coming from the same source. Thus, intelligent systems are data driven learning methodologies that seek to approximate optimal solutions of problems of high dimensionality. Since, in contrast, theory driven approaches give rise to precise specifications of the required algorithms for solving simplified models, the search for patterns replaces the search for reasons.

Our starting point is to examine the degree of randomness inhibited in the chosen EUR/USD, EUR/GBP, and EUR/USD time series. The strategy is to build econometric models in order to extract statistical dependencies within a time series that may be based on linear and/or nonlinear relationships. If such dependencies are significant, then the time series is not totally random since it contains deterministic components. These may be important indicators for the predictability of exchange rate returns. Second, we use time series analysis methods for building empirical models that will serve as benchmarks for the SVM specifications.

We represent the underlying problem of forecasting exchange rate ups and downs as a classification task. In supervised learning, it is assumed that a functional relationship is implicitly reflected within the input/output pairings. The estimate of the target function which is learnt or output by the learning algorithm is known as the solution of the learning problem. There exist many possible target functions that can separate the data. However, there is only one that maximizes the distance between itself and the nearest data point of each class. The problem of learning this *optimal* target function is formally represented by the binary classification problem (BCP). The BCP is a known problem in the field of computational geometry, which is the branch of computer science that studies algorithms for solving geometric problems. The input to a computational-geometry problem is typically a description of a set of geometric objects, such as a set of points, a set of line segments, or the vertices of a polygon. The output is a response to a query about these objects, or even a new geometric object. In its most general form, the case of whether two sets of points in general space can be separated by k hyperplanes, the BCP is known to be NP-complete, i.e., not solvable by a polynomial time algorithm.

In order to address the BCP, we propose the use of SVM, a state-of-the-art supervised learning system which, based on the laws of statistical learning theory [405], maps the input dataset via kernel function into a high-dimensional feature space in order to enable linear data classification. SVM has proven to be a principled and very powerful method that in the few years since its introduction has already outperformed many other systems in a variety of applications. The forecasting approach that we adopted is what we call a statistical or purely data driven approach that borrows from both fundamental and technical analysis principles. It is fundamental since it considers relationships between the exchange rate and other exogenous financial market variables. However, our approach also has a technical component: it is somewhat irrational in a financial context since it depends heavily on the concepts of statistical inference.

Ever, since the introduction of the SVM algorithm, the question of choosing the kernel has been considered as very important. This is largely due to the effect that the performance highly depends on data preprocessing and less on the linear classification algorithm to be used. How to efficiently find out which kernel is optimal for a given learning task is still a rather unexplored problem and subject to intense current research. We choose to take a pragmatic approach by comparing a range of kernels with regards to their effect on SVM performance. The argument is that even if a strong theoretical rational for selecting a kernel is developed, it would have to be validated using independent test sets on a large number of problems. The evaluation procedure is twofold. Out-of-sample forecasts are evaluated both statistically via confusion matrices and practically via trading simulations.

The results of Part III shed light on the existence of a particular kernel function which is able to represent properties of exchange rate returns generally well in high dimensional space. In particular, it is found that hyperbolic SVMs perform consistently well in terms of forecasting accuracy and trading performance via a simulated strategy. Moreover, we find that p-Gaussian SVMs, which have hardly been tested yet on real datasets, but in theory, have very interesting properties, perform reasonably well in predicting EUR/GBP and EUR/USD return directions. The results can be valuable for both practitioners, including institutional investors, private investors, risk managers, and researchers who will focus on SVM models and their technical improvements. We strongly believe, that given the shortcomings of economic and financial theories in explaining real-world market behaviors, modern methods of machine learning, pattern recognition, and empirical inference will play an essential role in coping with such complex environments.

2.3 Exchange Rate Hedging in a Simulation/Optimization Framework

In Part IV, it is implicitly assumed that a way to understand corporate hedging behavior is in the context of speculative motives that could arise from either overconfidence or informational asymmetries. A model for managing currency transaction risk is developed with the objective to find a possibly optimal combination of linear and nonlinear financial instruments to hedge currency risk over a planning period such that the expected utility at the planning horizon is maximized. We require the goal function to address the conflicting empirical finding that firms do like to try to anticipate events, but that they also cannot base risk management on second-guessing the market. For this purpose, we assume the fictitious firm to have different future expectations than those implied by derivative prices. When addressing these questions, one must recognize that traditional mean–variance or mean–quantile-based performance measures may be misleading if products with nonlinear payoff profiles such as options are used. In addition, the present knowledge that underlies the field of decision making is simple principles that define rationality in decision making and empirical facts about the cognitive limits that lead us not to decide rationally. As such, it has become a stylized fact that individuals perceive risk in a nonlinear fashion.

For these reasons, it is proposed to embed a mean–variance–skewness utility maximization framework with linear constraints in a single-period stochastic combinatorial optimization problem (SCOP) formulation in order to find optimal combinations for forward and European style straddle option contracts given preferences among objectives. The advantage of using SCOPs over deterministic combinatorial optimization problems (DCOPS) is that the solutions produced may be more easily and better adapted to practical situations where uncertainty cannot be neglected. The use of SCOPs instead of DCOPs naturally comes at a price: first, the objective function is typically much more computationally demanding. Second, for a practical application of SCOPs, there is the need to assess probability distributions from real data or subjectively, a task that is far from trivial. We show that the proposed SCOP is computationally hard and too difficult to be solved analytically in a reasonable amount of time. This suggests the use of approximate methods, which sacrifice the guarantee of finding optimal solutions for the sake of getting good solutions in a significantly reduced amount of time. Furthermore, computation of the objective function is nonconvex and nonsmooth. The solution space of the resulting optimization problem therefore becomes fairly complex as it exhibits multiple local extrema and discontinuities. In order to attack the given problem and derive near-optimal decisions within a reasonable amount of time, a simulation/optimization procedure is suggested. Simulation/optimization has become a large area of current research in informatics, and is a general expression for solving problems where one has to search for the settings of controllable decision variables that yield the maximum or minimum expected performance of a stochastic system as presented by a simulation model. Our approach to simulation/optimization addresses real-world complexity by incorporating the following technical features.

For modeling the exchange rate, a smooth transition nonlinear PPP reversion model is presented. Its key feature is a mathematical function which allows for smooth transition between exchange rate regimes, symmetric adjustment of the exchange rate for deviations above and below equilibrium, and the potential inclusion of a neutral corridor where the exchange rate does not mean revert but moves sideways. Apart from equilibrium which is given by PPP, only two parameters need to be estimated: the speed of PPP reversion and exchange rate volatility. We believe that the proposed exchange rate model is very attractive. It includes a long-run component (PPP), a medium-term component (adjustment speed), and a short-term component (volatility) which, depending on their estimated values, may overshadow each other and therefore cover a variety of phenomena observed in the real foreign exchange market.

Optimization of complex systems has been for many years limited to problems that could be formulated as mathematical programing models of linear, nonlinear and integer types. The best-known optimization tool is without a doubt linear programing. Linear programing solvers are designed to exploit the structure of a well defined and carefully studied problem. The disadvantage to the user is that in order to formulate the problem as linear program, simplifying assumptions and abstractions may be necessary. This leads to the popular dilemma of whether to find the optimal solution of models that do not necessarily represent the real system or to create a model that is a good abstraction of the real system but for which only very inferior solutions can be obtained. When dealing with the optimization of stochastic systems, obtaining optimal values for decision variables generally requires that one has to search for them in an iterative or ad hoc fashion. This involves running a simulation for an initial set of values, analyzing the results, changing one or more values, rerunning the simulation, and repeating the process until a satisfactory (optimal) solution is found. This process can be very tedious and time consuming and it is often not clear how to adjust the values from one simulation to the next. The area of metaheuristics arose with the goal of providing something better than simple searching strategies (e.g., grid search) by integrating high-level intelligent procedures and fast computer implementations with the ability to escape local optimal points. The specific metaheuristic we use is a variant of the scatter search algorithm, which has proven to be highly successful for a variety of known problems, such as the Vehicle Routing Problem, Tree Problems, Mixed Integer Programing, or Financial Product Design.

In order to show our simulation/optimization model's applicability in a practical context, a case study is presented, where a manufacturing company, located in the EU, sells its goods via a US-based subsidiary to the end-customer in the US. Since it is not clear what the EUR/USD spot exchange rate will be on future transaction dates, the subsidiary is exposed to foreign exchange transaction risk under the assumption that exposures are deterministic. We take the view that it is important to establish whether optimal risk management procedures offer a significant improvement over more ad hoc procedures. For the purpose of model validation, historical data backtesting is carried out and it is assessed whether the optimized mean–variance–skewness approach is able to outperform nonpredictive, fixed-weight strategies such as unitary spot, forward, and straddle, as well as a mixed strategy over time. We compare the alternative strategies in dynamic back-testing simulations using market data on a rolling horizon basis. The strategies are evaluated both in terms of their ex ante objective function values, as well as in terms of ex post development of net income.

We find in our experiments that scatter search is a search method that is both aggressive and robust. It is aggressive because it finds high-quality solutions early in the search. It is robust because it continues to improve upon the best solution when allowed to search longer. We find that our approach to hedging foreign exchange transaction risk adds value in terms of reducing risk and enhancing income. The optimized mean–variance–skewness strategy provides superior risk-return results in comparison to the passive strategies if earnings risk is perceived asymmetrically in terms of downside risk. Even with low levels of predictability, there is a substantial loss in opportunity when fixed-weight strategies (which assume no predictability) are implemented relative to the dynamic strategy that incorporates conditioning information. In particular, the pure forward strategy is found to have the lowest return per unit of earnings risk whereas the European style straddle strategy and the 1/3 strategy reveal similar risk-return characteristics. Interestingly, our research also contrasts the finding that currency forward contracts generally yield better results in comparison to options, since a passive straddle strategy would have yielded superior results compared to a forward strategy. Apart from our backtesting results, it is believed that the proposed simulation/optimization procedure for determining optimal solutions has important implications for policy making. Having easy access to relevant solutions makes it easier for policy makers to explore and experiment with the model while incorporating their own intuition before deciding on a final plan of action. Despite the many problems that economic forecasts from economic systems confront, these models offer a vehicle for understanding and learning from failures, as well as consolidating our growing knowledge of economic behavior.

Part II
Foreign Exchange Market Predictability

"The ultimate theory will place no limit on the complexity of systems that we can produce and it is in this complexity that I think the most important developments of the next millenium will be." (Stephen Hawking – Science in the next millennium, White House Millenium Council (2000))

Chapter 3
Equilibrium Relationships

3.1 Purchasing Power Parity Theorem

Purchasing power parity (PPP) theory states that in the long run, the exchange rate between the currencies of two countries should be equal to the ratio of the countries' price levels. PPP theory has a long history in economics, dating back several centuries, but the specific terminology was introduced in the years after World War I during the international policy debate concerning the appropriate level for nominal exchange rates among the major industrialized countries after the large-scale inflations during and after the war [69]. Since then, the idea of PPP has become embedded in how many international economists think about the world. For example, Dornbusch and Krugman [102] noted:

> Under the skin of any international economist lies a deep-seated belief in some variant of the PPP theory of the exchange rate.

Rogoff [342] expressed much the same sentiment:

> While few empirically literate economists take PPP seriously as a short-term proposition, most instinctively believe in some variant of purchasing power parity as an anchor for long-run real exchange rates.

Macroeconomic literature distinguishes between two notions of PPP: Absolute PPP and Relative PPP.

3.1.1 Absolute PPP

PPP follows from the law of one price, which states that in competitive markets, individual identical goods will sell for identical prices when valued in the same currency at the same time. The reason why the law of one price should hold is based on the idea of frictionless goods arbitraging: if prices were not identical, people could make a riskless profit by shipping the goods from locations where the price

C. Ullrich, *Forecasting and Hedging in the Foreign Exchange Markets,* Lecture Notes
in Economics and Mathematical Systems 623, DOI: 10.1007/978-3-642-00495-7_3,
© Springer-Verlag Berlin Heidelberg 2009

is low to locations where the price is high. If the same goods enter each market's basket used to construct the aggregate price level – and with the same weight – then the law of one price implies that PPP exchange rate should hold between the countries concerned.[1]

Absolute PPP holds when the purchasing power of a unit of currency is exactly equal in the domestic economy and in a foreign economy, once it is converted into foreign currency at the market exchange rate. Let P_t be the price of the standard commodity basket in domestic terms and P_t^* the price of the same basket in a foreign country. Formally, absolute PPP states that the real exchange rate R_t between two countries should be

$$R_t = S_t \frac{P_t^*}{P_t} = 1 \tag{3.1}$$

where S_t denotes the nominal exchange rate in domestic currency per unit of foreign currency. Taking the natural logarithm, absolute PPP can be stated as

$$\ln(S_t) + \ln\left(\frac{P_t^*}{P_t}\right) = 0 \tag{3.2}$$

or

$$s_t = p_t - p_t^* \tag{3.3}$$

Hence, the real exchange rate in its logarithmic form may be written as

$$r_t = s_t + p_t^* - p_t \tag{3.4}$$

3.1.2 Relative PPP

In a world where market participants have rational expectations, several factors might doubt the existence of absolute PPP. For instance, it is unrealistic to assume that all goods are identical, tradable, that there are no transportation costs, taxes, tariffs, restrictions of trade, or border effects and that competition is perfect. Thus, it has become common to test relative PPP, which holds if the percentage change in the exchange rate over a given period just offsets the difference in inflation rates in the countries concerned over the same period. If transportation costs and trade restrictions are assumed to be constant (c), relative PPP can be written as

---

[1] Nevertheless, the presence of any sort of tariffs, transport costs, and other nontariff barriers and duties would induce a violation of the law of one price. Also the assumption of perfect substitutability between goods across different countries is crucial for verifying the law of one price. In reality, however, product differentiation across countries creates a wedge between domestic and foreign prices of a product which is proportional to the freedom with which the good itself can be traded. An example often cited in the literature is the product differentiation of McDonald's hamburgers across countries. Examples for which the law of one price may be expected to hold are gold and other internationally traded commodities [342]. For a discussion on the empirical evidence of the law of one price see [357].

$$\ln(S_t) = c + \ln\left(\frac{P_t}{P_t^*}\right) \qquad (3.5)$$

After taking first differences, price level changes determine the exchange rate development

$$\Delta\ln(S_t) = \Delta\ln\left(\frac{P_t}{P_t^*}\right) \qquad (3.6)$$

The expected exchange rate change thus equals the difference in inflation of the respective countries. Thus, if the EU price level rose 10% and the US price level rose 5%, the Euro would depreciate 5%, offsetting the higher EU inflation and leaving the relative purchasing power of the two currencies unchanged. If absolute PPP holds, then relative PPP must also hold. However, if relative PPP holds, then absolute PPP does not necessarily hold, since it is possible that common changes in nominal exchange rates are happening at different levels of purchasing power.

3.1.3 Empirical Evidence

The validity of absolute and relative PPP and the properties of PPP deviations have been the subject of an ongoing controversy in economic and econometric literature. Until the 1970s, early empirical studies have examined by ordinary least squares (OLS) regression analysis, whether there is a significant linear relationship between relative prices and the exchange rate. In general, such tests led to a rejection of the PPP hypothesis except for countries exhibiting high inflation [136] which suggested that PPP may represent an important benchmark in long-run exchange rate modeling. However, the OLS method suffers from econometric weaknesses since it ignores the econometric requirement that residuals of the estimated regression equation must be stationary. If stationarity is not given, the use of nonstationary data can lead to a spurious regression, i.e., nonsense correlation [172], which makes it impossible to validly undertake conventional OLS based statistical inference procedures. Otherwise, if linear regression residuals are stationary, then although a strong long-run linear relationship exists between exchange rates and relative prices, it is still invalid to undertake conventional statistical inference because of the bias present in the estimated standard errors [26, 116]. Thus, although some researchers could not reject the argument that PPP deviations are nonstationary and therefore permanent (for instance [6, 133, 343]), it does not necessarily mean that these findings must be accepted.

In 1987, the concept of cointegration emerged [116, 209, 210] which seemed to be an ideal approach to testing for PPP. A cointegrated system requires that any two nonstationary series which are found to be integrated of the same order are cointegrated if a linear combination of the two exists which is itself stationary. If this is the case, then the nonstationarity of one series exactly offsets the nonstationarity of the other series and a long-run relationship is established between the two variables. Testing for *no cointegration* led to mixed results. While some studies in the late 1990s resulted in reports about the absence of significant mean reversion of the real exchange rate for the recent floating experience [276], other applied work on long-run PPP among the major industrialized economies has been more favorable towards the long-run PPP hypothesis (e.g., [72, 73, 82, 228]).

One important reason for the fact that statistical tests of the 1980s to examine the long-run stability of the real exchange rate often failed was given by [133, 134] (followed by [142, 268]) who argued that these tests are not powerful enough. This *power problem* cannot simply be solved by increasing the sample size from monthly to daily data, since increasing the amount of detail concerning short-run movements can only give more information about short-run as opposed to long-run behavior [367]. In order to get more information about the long-run behavior of a particular real exchange rate, one approach is to use more years of data. However, long periods of data may span different exchange rate regimes, and therefore exhibit regime switches or even structural breaks. In an early study in this spirit, using annual data from 1869 to 1984 for the USD/GBP real exchange rate, [133] was able to reject the hypothesis of a random walk at the 5% level. Similar results to Frankel's were obtained by [160], and [268] which were also unable to detect any significant evidence of a structural break between the pre- and post-Bretton Woods period. Taylor [382] extended the long-run analysis to a set of 20 countries over the period 1870–1996 and also finds support for PPP and coefficients that are stable in the long run. A different approach to providing a convincing test of real exchange rate stability, while limiting the time frame to the post-Bretton Woods period, is to increase the panel by using more countries. By increasing the amount of information employed in the tests across exchange rates, the power of the test should be increased. Abuaf and Jorion [2] examined a system of ten first-order autoregressive regressions for real dollar exchange rates over the period 1973–1987, where the autocorrelation coefficient is constrained to be the same in every case. Their results indicate a marginal rejection of the null hypothesis of joint nonmean reversion at conventional significance levels, which they interpret as evidence in favor of long-run PPP. In a more recent panel data study on Euro exchange rates, [267] found strong rejections of the random walk hypothesis. Taylor and Sarno [384], however, issued an important warning in interpreting these findings. The tests typically applied in these panel-data studies test the null hypothesis that none of the real exchange rates under consideration are mean reverting. If this null hypothesis is rejected, then the most that can be inferred is that at least one of the rates is mean reverting. However, researchers tended to draw a much stronger inference that all of the real exchange rates were mean reverting and this broader inference is not valid. Some researchers have sought to remedy this shortcoming by designing alternative tests. For example, [384] suggested testing the hypothesis that at least one of the real exchange rates is nonmean reverting, rejection of which would indeed imply that they are all mean reverting. However, such alternative tests are generally less powerful, so that their application has not led to clear-cut conclusions [356, 384].

Another important stage in PPP literature was initialized in the mid-1980s when [199] and others began to notice that even the studies that were interpreted as supporting the thesis that PPP holds in the long run, suggested that the speed at which the real exchange rates adjusted to PPP was extremely slow. This involved the computation of the so-called half-life of shocks to the real exchange rate, which describes how long it would take for the effect of a shock to die out by 50%. By including both long-span investigations and panel unit root tests of long-run PPP, [342] finds that estimated half-lives of adjustment mostly tend to fall into the range of 3–5 years. Some more recent studies have argued that half-lives are even longer if estimation is not based on OLS (for example, [71, 302]). The apparently very slow speed of adjustment of real exchange rates paired with further criticism on the insufficiency of traditional linear statistical testing procedures led to an interesting body of research which argues that failure to accept the long-run PPP hypothesis could also be due to a different reason: the existence of nonlinear dynamics in the real exchange rate. If nonlinear dynamics existed then the exchange rate would become increasingly mean reverting with the size of the deviation from equilibrium [106, 227].

To summarize, if exchange rates do tend to converge to PPP, economists have – at least so far – had a hard time presenting strong evidence to support the claim. The difficulty for validating PPP empirically has been captured in two PPP puzzles:

1. The *disconnect puzzle* states that the difficulty of detecting evidence for long-run PPP suggests that the exchange rate is disconnected from PPP and therefore violates the PPP theorem [312, 385, 386].
2. The *excess volatility puzzle* states that the enormous short-term volatility of real exchange rates is contradicting to the extremely slow rate of adjustment ([342], p. 647).

Current consensus view on PPP research supports the hypothesis that only relative PPP seems to hold in the long run. Structural and transitory effects influence the real exchange rate permanently and make absolute PPP obsolete. PPP is generally considered to be a meaningful element of macroeconomics for an open economy in the long-run, at least as a benchmark for over- or undervaluation of a currency. However, it offers no explanation for short term exchange rate variation. A second stylized fact is that PPP seems to hold better for countries with relatively high rates of inflation and underdeveloped capital markets [385].

3.1.4 Explanations for Deviations from PPP

The empirical failure to validate PPP has been subject to various explanations. From an empirical perspective it has been argued that it is difficult to find a price index that accurately measures the inflation rate for the countries being studied [136, 254]. The problem of simultaneous determination of both price and the foreign exchange rate is noted by [178]. Pippenger [325] claims that one obstacle to finding empirical support for PPP may be due to the statistical procedures applied.

From a theoretical perspective, macroeconomic literature has explained long run departures from PPP by the Harrod–Balassa–Samuelson hypothesis which depends on inter-country differences in the relative productivity of the tradable and non-tradable sectors [23, 184, 353]. Another important explanation addressing short-run departures from PPP was given by Dornbusch's theory on how expansionary monetary policy may lead to overshooting nominal exchange rates [102]. In short, prices on goods are assumed to be *sticky* in the short-run while the exchange rate as a financial market price quickly adjusts to disturbances initiated by monetary policy, i.e., increases or decreases in the money supply. In the short-run, the exchange rate is therefore determined by the supply and demand of financial assets. The portfolio balance theory [101] extends Dornbusch's sticky price monetary model by incorporating another important macroeconomic variable: the current account balance. According to this theory, a monetary shock (e.g., increase in money supply) is expected to affect (increase) prices which affects net exports and hence influences the current account balance. In turn, this will affect the level of wealth which feeds back into the asset market, affecting the exchange rate during the adjustment to long-run equilibrium. The general tenor on these traditional exchange rate theories is that, although plausible, empirical work has neither favored a particular theory nor produced related models that are sufficiently statistically satisfactory to be considered reliable and robust (see references in [357], Chap. 4). This shortcoming has led to a rapid development of new open-economy models which avoid the formal simplicity of the traditional models by offering a more rigorous analytical foundation through fully specified microfoundations (see for instance [311]). In this spirit, [66] showed that in a small open economy with sticky prices (i.e., prices exhibiting resistance to change) and a nontraded sector, small monetary shocks can generate high levels of exchange rate volatility. This could be an explanation for the second PPP puzzle. However, the problem with open-economy models is that they are less universal and robust. In addition they require assumptions which are difficult to test (see references in [357], Chap. 5).

Econometric literature distinguishes between permanent (structural) or temporary (transitory) shocks to the exchange rate fashion by focusing on the employment of vector autoregression (VAR) analyses, an econometric technique which intends to capture the evolution and the interdependencies between multiple time series [42]. Overall, literature on identifying the source of structural and transitory shocks driving exchange rates has provided mixed results. While both nominal shocks (e.g., monetary shocks) and real shocks (initiated by real factors such as real income, factor endowment, productivity levels, interest rates, etc.) were shown to be sources of exchange rate volatility leading to departures from PPP, their relative importance varies across studies (see for instance [88, 244, 340]).

A promising avenue of research addressing both PPP puzzles at once investigates the role of nonlinearities in real exchange rate dynamics. If adjustment to long-run equilibrium occurs in a nonlinear fashion, then the speed of adjustment varies depending on the magnitude of disequilibrium. Thus, while small shocks to the real exchange rate around equilibrium will be highly persistent, larger shocks mean-revert much faster. This behavior may be motivated economically by market frictions such as

- Barriers of international trade including transport costs , tariffs, nontariff barriers [128, 342]
- Transaction costs [33, 106, 419]
- Heterogeneity of opinion [227]
- Central bank interventions [368, 383]

The nonlinear nature of the adjustment process has been investigated in terms of the threshold autoregressive (TAR) model [313, 393] the smooth transition autoregressive (STAR) model [174, 292, 386] and Markov switching models [232, 383].

3.2 Interest Rate Parity (IRP) Theorem

When an asset market is taken into account, then equilibrium theory follows the essence of international manifestations of the law of one price. The interest rate parity (IRP) theorem is an arbitrage condition which illustrates the idea that, in the absence of market imperfections, risk-adjusted expected real returns on financial assets will be the same in foreign markets as in domestic markets. Hence, returns generated from borrowing in domestic currency, exchanging that currency for foreign currency and investing in interest-bearing foreign currency denominated assets, while simultaneously purchasing forward contracts to convert the foreign currency back at the end of the investment period should be equal to the returns from purchasing and holding similar interest-bearing instruments of the domestic currency. Let $F_{t,k}$ denote the k-period forward rate, i.e., the rate agreed now for an exchange of currencies k periods ahead. Let further be $i_{t,k}$ and $i_{t,k}^*$ the nominal interest rates available on similar domestic and foreign securities respectively with k periods to maturity. Formally, IRP states that the ratio of the forward and spot exchange rates should be

$$\frac{F_{t,k}}{S_t} = \frac{1 + i_{t,k}^*}{1 + i_{t,k}} \tag{3.7}$$

IRP thus holds, if the ratio of the forward and spot exchange rates equals the ratio of foreign and domestic nominal interest rates. If (3.7) does not hold, investors can theoretically arbitrage and make risk-free returns. Academic literature commonly distinguishes between two versions of the identity: covered interest rate parity (CIP) and uncovered interest rate parity (UIP).

3.2.1 Covered Interest Rate Parity (CIP)

CIP is an arbitrage condition which postulates, that the interest rate difference between two countries' currencies, is equal to the percentage difference between the forward exchange rate and the spot exchange rate. This requires the assumptions that financial assets are perfectly mobile and similarly risky. Taking the natural logarithm in (3.7) and using the approximation $\ln(1+x) \approx x$, CIP holds if

$$f_{t,k} - s_t = i_{t,k} - i_{t,k}^* \tag{3.8}$$

where $f_{t,k}$ is the logarithm of the k-period forward rate, s_t denotes the logarithm of the spot exchange rate at time t, $i_{t,k}$ and $i_{t,k}^*$ are the nominal interest rates available on similar domestic and foreign securities, respectively, with k periods to maturity.

A direct implication of (3.8) is, that when the domestic interest rate is lower than the foreign interest rate $i_{t,k} - i_{t,k}^* < 0$ the forward price of the foreign currency will be below the spot price $f_{t,k} - s_t < 0$, i.e., the foreign currency is selling forward at a premium. Conversely, if domestic interest rate is higher than the foreign interest rate $i_{t,k} - i_{t,k}^* > 0$, the forward price of the foreign currency will be above the spot price $f_{t,k} - s_t > 0$, i.e., the foreign currency is selling forward at a discount. Hence, the national interest rate difference should be equal to, but opposite in sign, to the forward rate premium or discount for the foreign currency.

3.2.2 Uncovered Covered Interest Rate Parity (UIP)

UIP postulates that the return on domestic currency deposit should be equal to the expected return from converting the domestic currency into foreign currency, investing it in a foreign currency denominated asset and then converting the proceeds back into the domestic currency at the future expected exchange rate. Formally, the interest rate difference between two countries' currencies is equal to the percentage difference between the expected exchange rate and the spot exchange rate

$$E(s_{t+k}) - s_t = i_{t,k} - i_{t,k}^* \tag{3.9}$$

where $E(s_{t+k})$ denotes the market expectation for s_t in k periods based on information at t. Unlike CIP, UIP is not an arbitrage condition since the expected exchange rate $E(s_{t+k})$ is unknown at time t and therefore nonzero deviations from UIP do not necessarily imply the existence of arbitrage profits due to the foreign exchange risk associated with future exchange rate movements.

3.2.3 Empirical Evidence

In reality, spot and forward markets are not always in a state of equilibrium as described by IRP. As a consequence, arbitrage opportunities exist. Arbitrageurs who recognize a state of disequilibrium can borrow in the currencies exhibiting relatively low interest rates and convert the proceeds into currencies which offer higher interest rates. In case that the investor is hedged against any risk of the currency deviating by selling the currency forward, this is known as covered interest arbitrage (CIA). If the investor does not sell the currency forward, thus remaining exposed to the risk of the currency deviating, this is known as uncovered interest arbitrage (UIA). The motivation for entering into UIA is seeking to profit from expected changes in the exchange rate, i.e., rational speculation.

CIA can be tested in two ways. The first approach relies on computing the actual deviations from IRP to see if they differ significantly from zero. The significance of departures from CIP is often defined with respect to a neutral band, which is determined by transaction costs. Frenkel and Levich [138] calculated a band around the IRP line within which no arbitrage is possible. Using weekly data for the period 1962–1975 and for three sub-periods, they found that around 80% of observations are within the neutral band. They conclude that, after allowing for transaction costs and ensuring that the arbitraged assets are comparable, CIA does not seem to entail unexploited opportunities for profit. This finding is confirmed by [79] who shows that deviations from CIP should be no greater than the minimum transaction costs in one of three markets. On the basis of analysis of data for five major currencies against the US dollar, Clinton finds that the neutral band should be within +0.06% per annum from parity and that although the hypothesis of zero profitable deviations from parity can be rejected,

> empirically, profitable trading opportunities are neither large enough nor long-lived enough to yield a flow of excess returns

over time to any factor. However, there also exists contrary evidence in favor of arbitrage opportunities in the foreign exchange markets. By questioning the quality of the data used by [138], various researchers have often arrived at different conclusions. Taylor [380, 381]), for instance, argues that in order to provide a proper test of CIP it is important to have data on the appropriate exchange rate and interest rates recorded at the same instant in time at which a dealer could have dealt. For this purpose, he uses high-quality, high-frequency, contemporaneously sampled data for spot and forward exchange rates and corresponding interest rates for a number of maturities and makes allowance for bid-offer spreads and brokerage costs in his calculations. He finds that there are few profitable violations of CIP, even during periods of market uncertainty and turbulence. In a more recent study, [25] examined the dynamics of deviations from CIP using daily data on the GBP/USD spot and forward exchange rates and interest rates over the period January 1974 to September 1993. They find, as opposed to [138] that a substantial number of instances in the sample in which the CIP condition exceeds the transaction costs band, imply arbitrage profit opportunities. In addition, [283] identified significant arbitrage opportunities from ten markets over a 12-day period. Some of these opportunities were even observed to be persistent for a long time. Persistent interest rate arbitrage opportunities were also revealed by [79] and [1]. An alternative approach for testing the validity of CIP is to estimate the regression coefficients of

$$f_{t,k} - s_t = \alpha + \beta (i_t - i_t^*) + \varepsilon_t \qquad (3.10)$$

which are required to be $(\alpha, \beta) = (0, 1)$ and an uncorrelated error. Equation (3.10) has been tested by a number of researchers for a variety of currencies and time periods (see, e.g., the early study by [59]). The main conclusion to be drawn from this line of research is that, CIP is supported. Although there are significant deviations from the condition $\alpha = 0$ which might reflect the existence of nonzero transaction costs, the estimates of b usually differ insignificantly from 1.

Assuming that CIP holds and market participants have rational expectations, UIP implies that forward premium should be an *unbiased predictor* of the expected change in the spot exchange rate. UIP can therefore be tested by estimating a regression of the form:

$$s_{t+k} - s_t = \alpha + \beta (f_t^k - s_t) + \varepsilon_{t+k} \tag{3.11}$$

Under UIP, $(\alpha, \beta) = (0, 1)$ and the rational expectations forecast error ε_{t+k} must be uncorrelated with information available at time t [125]. Empirical studies generally report results which reject UIP (e.g., see the references in the surveys of [193, 263, 357]) suggesting that it would have been possible to make speculative gains in certain markets over certain periods of time. One reason is the empirical finding that the spot exchange rate change on mostly free floating nominal exchange rates up until 1990s appears to be negatively correlated $(b < 0)$ with the lagged forward premium or forward discount [143]. This observation implies that the more the foreign currency is at a premium in the forward market, the less the home currency is predicted to depreciate, which has been referred to as the forward premium anomaly or forward premium puzzle [20]. Fama [125] finds further but different empirical evidence against UIP: forward rates are not unbiased but biased predictors of future spot rates because a nonnegative interest rate differential would, on average, result in an appreciating currency for the high interest rate country. This observation has been referred to as *forward bias puzzle* ([263]). A simple rule to capitalize on a forward rate bias would be to enter into a forward hedge whenever the foreign currency is offered at a premium and never hedge when it is at a discount ([233]).

3.2.4 Explanations for Deviations from IRP

Different theoretical explanations have been proposed to explain empirical deviations from CIP and UIP and the related rejection of the efficient markets hypothesis. For instance, the existences of a risk premium due to risk-averse market participants is one prominent attempt to explain the UIP forward bias puzzle. If foreign exchange market participants are risk-averse, the UIP condition may be distorted by a risk premium $\rho_{t,k}$, because agents demand a higher rate of return than the interest differential in return for the risk of holding foreign currency [125]. The difference between the forward exchange rate and the spot exchange rate has always been used as the measure of the risk premium that the marginal investor would be willing to pay in order to reduce his exposure to exchange risk. The message which emerges from the empirical analysis of risk-premium models is that it is hard to explain excess returns in forward foreign exchange by an appeal to risk premiums alone [30, 31, 269, 275]. Chinn [74] and McCallum [285] therefore propose to link the behavior of UIP deviations with monetary policy reaction functions.

A second explanation for the forward bias puzzle is that there is a failure of the rational expectations component underlying the notion of market efficiency. Literature identifies several possible explanations for volatile expectations or departures from rational expectations that generate nonzero and potentially predictable excess returns even when agents are risk neutral: for instance, learning about regime shifts [261, 262] or about fundamentals, such as learning about the interest rate process [169], the *peso problem* originally suggested by [341], or inefficient information processing [37]. This literature is well covered in several surveys (e.g., [114, 263, 357]).

Third, it could be possible that both the risk-aversion of market participants and a departure from the rational expectations hypothesis are responsible for the rejection of the efficient markets hypothesis. Important contributions in this area include the work by [135] and [141].

A more general argument referring to the rejection of both CIP and UIP is that the relationship between exchange rate and interest rate is not a linear, but a nonlinear one. This could be for reasons such as transaction costs [24, 195, 363], central bank intervention [277], limits to speculation ([270], pp. 206–220), or information costs ([21], p. 50).

Chapter 4
Market Efficiency Concepts

4.1 Informational Efficiency

The theoretical concepts of informational and speculative market efficiency are closely related to the IRP Theorem. According to [124] and [126], a financial market is said to be (informationally) efficient, if prices in that market fully reflect all the available and relevant information. The intuition is, that, if the market processes that information immediately, price changes can only be caused by the arrival of new information. However, since future information cannot be predicted, it is also impossible to predict future price changes. Depending on the information set available, there are different forms of the market efficiency hypothesis (MEH):[1]

- *Weak-form efficiency.* No investor can earn excess returns by developing trading rules based on historical price or return information. In other words, the information in past prices or returns is not useful or relevant in achieving excess returns.
- *Semistrong-form efficiency.* No investor can earn excess returns by developing trading rules based on publicly available information. Examples of publicly available information are annual reports of companies, investment advisory data, or ticker tape information on TV.
- *Strong-form efficiency.* No investor can earn excess returns using any information, whether publicly available or not. In other words, the information set contains all information, including private or insider information.

If applied to the foreign exchange markets, informational efficiency means that the current spot rate has to reflect all currently available information. This has one important implication: if expectations about the future exchange rate are rational [304] they should all be incorporated in the forward rate. Hence, if the efficient market

---

[1] Latham [245] and Rubinstein [349] have extended the definition of market efficiency by pointing out that it is possible that people might disagree about the implications of a piece of information so that some buy an asset and others sell in such a way that the market price remains unaffected. As a consequence, they require not only that there be no price change but also that there be no transactions. Still, theoretical and applied literature focuses on the efficiency definition as given by [124] and [126], which will provide the foundation for the following considerations, too.

C. Ullrich, *Forecasting and Hedging in the Foreign Exchange Markets,* Lecture Notes in Economics and Mathematical Systems 623, DOI: 10.1007/978-3-642-00495-7_4,
© Springer-Verlag Berlin Heidelberg 2009

hypothesis holds true, the forward rate must be an "unbiased predictor" for the expected exchange rate ([137]). When the forward rate is termed an unbiased predictor, then it over or underestimates the future spot rate with relatively equal frequency and amount. It therefore misses the spot rate in a regular and orderly manner such that the sum of the errors equals zero. As a direct consequence, it is not possible to make arbitrage profits since active investment agents will exploit any arbitrage opportunity in a financial market and thus will deplete it as soon as it may arise. Empirical tests for a bias in forward rates have been based on testing the two principles of CIA and UIA which have been referred to in Sect. 3.2.

4.2 Speculative Efficiency

A different concept of efficiency is *speculative efficiency* ([37, 54], p. 67), which, in contrast to informational efficiency does not imply that expectations about the future exchange rate must be rational and should all be incorporated in the forward rate. Instead, departures from the rational expectations hypothesis are allowed. The supply of speculative funds is infinitely elastic at the forward price, i.e., equal to the expected future spot price. The expected future spot price is a market price determined as the solution to the underlying irrational expectations model. The speculative efficiency hypothesis is tested by formulating a trading strategy and then calculating its profitability in simulated trading.

Chapter 5
Views from Complexity Theory

5.1 Introduction

The behavior of exchange rates is puzzling and hard to explain. While all of the potential explanations for deviations from equilibrium conditions appear fairly reasonable and have theoretical merit, PPP and IRP (forward premium anomaly) puzzles
have not yet been convincingly explained and continue to puzzle the international
economics and finance profession. However, this inexplicability has nothing to do
with economic and financial theorists being not capable enough but rather with the
complex dynamics that drive exchange rates and therefore, with the nature of complexity itself. Shiller [366] notes that all economic models have one major flaw:
a gross oversimplification that is based on the assumption that economic agents
know the true state of economic structure and make rational decisions for their consumption and investment. In a similar way, [351] argues that the inexplicability of
such complex systems may have to do with the existence of fundamental limits to
knowledge:

> This would suggest that financial theory cannot hope to provide much more than a dis-
> joint collection of highly simplified "toy models" which are either sufficiently stylized to be
> solved analytically, or are simple enough to be approximately solved on digital computers.
> In fact, conventional economic and financial theory does not choose to study the unfolding
> of the patterns its agents create. By assuming behavioral equilibrium, it rather simplifies its
> questions in order to seek analytical solutions. For example, standard theories of financial
> markets assume rational expectations and ask: what forecasts (or expectations) are consis-
> tent with – are on average validated by – the prices these forecasts and expectations together
> create. But it does not account for actual market "anomalies" such as unexpected price bub-
> bles and crashes, random periods of high and low volatility, and the use of heavy technical
> trading.

Hence, it is legitimate to ask whether economic and financial theory can provide
sufficient explanations for reality.

Complex systems theory with regards to markets has its origins in the classical
eighteenth century political economy of the Scottish Enlightenment which realized
that order in market systems is spontaneous or emergent in that it is the result of

C. Ullrich, *Forecasting and Hedging in the Foreign Exchange Markets,* Lecture Notes
in Economics and Mathematical Systems 623, DOI: 10.1007/978-3-642-00495-7_5,
© Springer-Verlag Berlin Heidelberg 2009

human action and not the execution of human design. This early observation, well known also from the Adam Smith metaphor of the invisible hand, premises a disjunction between system wide outputs, the modeling capabilities of individuals at a micro level, and the distinct absence of an external organizing force. It is therefore postulated that macroscopic properties cannot be formally or analytically deduced from the properties of its parts. This distinguishes the sciences of complexity theory from traditional economic and financial science which relies on deductive formalistic and analytical methods.

It was not until the twentieth century with two significant developments in the foundations of mathematics and advances in computer technology that complex adaptive systems could be investigated and formulated. First, [168, 329, 398] established logical impossibility limits to formalistic calculation or deductive methods providing first results on incompleteness, algorithmically unsolvable problems, and the notion of computational complexity. Usually problems solvable by a deterministic algorithm in polynomial time (i.e., the number of elementary computation steps is polynomial in the size of the input) are considered to be "good" solutions. On the contrary, a problem is considered as computationally intractable, if it is so hard that no polynomial time algorithm can possibly solve it.

Definition 5.1. A polynomial time algorithm is defined to be one whose time complexity function is $O(p(n))$ for some polynomial function p, where n is used to denote the input length. Any algorithm whose time complexity function cannot be so bounded is called an exponential time algorithm (although this definition includes certain nonpolynomial time complexity functions which are not normally regarded as exponential functions).

Problems that can be solved by a polynomial time algorithm are tractable in a sense that they coincide with those that can be realistically solved by computers, and it is held that every practical and efficient algorithm can be rendered as a polynomial time bounded Turing machine (see [152, 320]). On the other hand, one of the earliest and most popular intractability results are the classical undecidability results of [398], who proved that it is impossible to specify any algorithm which given an arbitrary computer program and an arbitrary input to that program, can decide whether or not the program will eventually halt when applied to that input.

Parallel efforts to the search of powerful methods for proving problems intractable have been made by focusing on learning more about the ways in which various problems are interrelated with respect to their difficulty. The theory of *NP*-completeness has introduced a variety of complexity classes with respect to certain problem formulations [152]. There are two important classes of problems which can be met in many areas of practical application.

Definition 5.2. The complexity class P is the set of decision problems that can be solved by a deterministic algorithm in polynomial time in the size of the input.

This class corresponds to an intuitive idea of the problems which can be effectively solved in the worst cases.

Definition 5.3. The complexity class NP is the set of decision problems that can be solved by a nondeterministic algorithm in polynomial time.

A nondeterministic algorithm is composed of two stages. The first stage is a guessing stage where, given a problem instance I, a structure S is guessed. The second stage is a checking stage, where given I and S as inputs, the algorithm behaves in a deterministic manner and halts with either an answer "yes" or "no." The class *NP* contains many problems that people would like to be able to solve effectively. Most of the apparently intractable problems encountered in practice, when phrased as decision problems, belong to this class.

Definition 5.4. A problem is *NP*-hard, if each problem in *NP* is reducible to it in polynomial time.

Definition 5.5. A problem is *NP*-complete, if it is *NP*-hard and inside the class *NP* of problems.

Hence, the *NP*-complete problems are the hardest problems among all those that can be solved by a nondeterministic algorithm in polynomial time and until today there has not been any solution for such a problem by a deterministic algorithm in polynomial time concerning the size of the input. The question of whether P is the same set as *NP* has been called the most important open question in theoretical computer science.

Apart from computational complexity, the second significant methodological development is *dynamic complexity* which considers the use of computer based artificial environments to simulate dynamics from large numbers of interacting agents with varying levels of computational and adaptive intelligence [14, 22, 194]. The seminal work of [358] is one of the earliest examples of the use of computer simulation to demonstrate how simple microbehavioral rules result in a self-organized macro outcome, an undesirable one of racial segregation, which could not have been deduced from the initial rules. More recently, a number of economists and physicists have got involved in the new computational agent based modeling in economics giving rise to very interesting and important new streams of literature called *adaptive computational economics* [346] and *econophysics* [211].

In the following, it is not our purpose to provide an extensive summary on these lively disciplines of research. Instead, we want to create awareness in the economic and financial community, that both the computational and dynamic complexity of economic calculations are proveably fundamental sources for real-world phenomena such as disequilibrium and market inefficiencies. We think that this perspective is important as it sheds some light on the feasibility of economic concepts such as the equilibrium concept. Thus, if a Turing machine cannot efficiently compute equilibrium then neither can a market. As we will show in the following, there are provable results on the impossibility for market agents to behave according to traditional economic concepts such as equilibrium, rationality, and efficiency.

5.2 Calculating Fixed Point Market Equilibrium

5.2.1 Computational Complexity of Centralized Equilibrium

The existence of equilibrium does not imply that it can actually be achieved. In the classic Walrasian model [413], the price formation mechanism (*Tâtonnement*) is simple and has the following properties:

- Agents truthfully reveal their preferences
- No trading takes place before the market-clearing price vector is announced
- All agents trade at exactly the same prices
- Final prices and allocations are completely determined from agent preferences and endowments
- Agent behavior is very simple, involving nothing more than truthful reporting of demands at announced prices.

Unfortunately, the job of the Walrasian *auctioneer* as the central entity to compute the equilibrium price is extremely hard. Equilibrium computation requires a mechanism for converging to a fixed point in a finite length of time, using a bounded amount of resources.

Definition 5.6. A fixed point of a function is a point that is mapped to itself by the function. That is to say, x is a fixed point of the function f, if and only if $f(x) = x$.

Without such a mechanism there is little reason to believe that a fixed point would ever be observed. Fixed-point theorems were introduced into economic theory by von Neuman in his work on the input–output model [409, 410]. Since then, many domains of economic theory have come to depend on fixed point theorems to prove the existence of equilibria, notably general equilibrium theory, but also Nash equilibria in game theory [308]. For a recent perspective on this, see [153]. The Brouwer fixed point theorem was one of the early achievements of algebraic topology, and provided the basis of more general fixed point theorems, such as Brouwer and Kakutani's fixed point theorem.

Theorem 5.1. *Let B^n be the n-dimensional closed unit ball and let $f : B^n \longrightarrow B^n$ be a continuous function. Then f has a fixed point: for some $x \in B^n$, $f(x) = x$.*

The proof of Brouwer's fixed point theorem assumes that either a triangle is being mapped onto itself, or another two-dimensional space is being mapped onto a triangle. Then a series of triangulations of decreasing size is constructed, and Sperner's Lemma is used to help conclude that the convergence of the series proves the existence of a fixed point.

Lemma 5.1. (Sperner's Lemma). *Suppose the interior of a triangle is triangulated (that is, divided up internally into small triangles). The vertices of the triangle are colored red, green, and blue, respectively. All other vertices, where lines meet inside or around the outside edges of the triangle, are also colored red, green, or blue with the restriction that no edge of the main triangle contains all three colors. Let T_C*

be the number of small triangles whose vertices are colored red, green, and blue in clockwise order; and let T_A be the number of small triangles whose vertices are colored red, green, and blue in anticlockwise order. Then

$$|T_C - T_A| = 1 \qquad (5.1)$$

In particular, the total number of small red–green–blue triangles must be odd, and is certainly never zero.

One famous real-world example of this theorem is stirring a glass of water. When the molecules stop moving, there will be one molecule that is in the same place it was before the stirring began. Another example involves two identical pieces of paper. If one of the pieces is crumpled and set on top of the other, there will be one point on that paper that lies directly above the corresponding point of the uncrumple paper.

Algorithms for computing Brouwer fixed points fall into a complexity class that makes them among the hardest problems in all of computer science. For instance, the lower bound for worst-case computation of Brouwer fixed points is exponential in the dimension of the problem [192] where the dimension is the size of the commodity space in the Arrow–Debreu version of general economic equilibrium [13]. Furthermore, it has recently been shown that the computational complexity of Brouwer and Kakutani fixed points are closely related to the complexity of the parity argument, the connection between the two being Sperner's Lemma [319]. The constructive problem arising from the application of Sperner's Lemma to the Brouwer and Kakutani fixed points of the Walrasian equilibrium model is that there are no polynomial time algorithms for the general case with nonlinear utility functions. Such results can only be obtained for linear utility functions.

5.2.2 Computational Complexity of Decentralized Equilibrium

It has been argued that the Walrasian model of exchange is problematic for several reasons. Recent results on the computational complexity of Brouwer and Kakutani fixed points suggest that real markets cannot possibly operate according to the Walrasian model. Furthermore, the Walrasian market model has no empirical underpinnings (e.g., [185], p. 55) and therefore does not present a reasonable picture of how an exchange economy works. The aggregation of information into the net demand functions of the Walrasian model has long been known to be excessive. Hayek [186] notes

> We cannot expect that this problem will be solved by first communicating all this information to a central board which, after integrating all knowledge, issues its orders [...]. The problem is to show how a solution is produced by interaction of people each of whom has partial knowledge.

In addition, there are a variety of non-Walrasian exchange mechanisms that yield equilibrium allocations that are Pareto optimal. These mechanisms are radically more decentralized than the Walrasian one, with its single uniform price vector, and display a more realistic picture of real economic processes. Rust ([351]) recommends using massively parallel computers and decentralized algorithms that allow competitive equilibria to arise as *emergent computations*.

> The reason why large scale computable general equilibrium problems are difficult for economists to solve is that they are using the wrong hardware and software. Economists should design their computations to mimic the real economy, using massively parallel computers and decentralized algorithms that allow competitive equilibria to arise as "emergent computations" [...]. The most promising way for economists to avoid the computational burdens associated with solving realistic large scale general equilibrium models is to adopt an "agent-based" modeling strategy where equilibrium prices and quantities emerge endogenously from the decentralized interactions of agents.

By using a decentralized k-lateral exchange model with local price formation mechanisms which studies the performance of market systems as a function of their scale, i.e., the number of agents in the marketplace and the number of commodities being traded, [19] analytically derives complexity properties for economies as large as a million agents and 20,000 commodities per agent. These properties are better (i.e., less complex) than those of Walrasian exchange models. Due to the fact that polynomial complexity is a prerequisite of real world equilibrium price formation mechanisms, [19] decentralized systems are the more plausible computational device.

5.2.3 Adaptive/Inductive Learning of Rational Expectations Equilibria

A major part of the coordination processes in markets relies on the identification and calculation of fixed point mappings of global signals such as equilibrium prices. By forming rational expectations of fixed points of market equilibrium prices, economic agents are meant to coordinate their activity and iron out inconsistent expectations. Spear [374] was the first to show that the problem of identifying a set of fixed points for market equilibrium price functions is algorithmically unsolvable.

The absence of a unique decision procedure or a deductive means by which to select appropriate forecast models/functions has been intuitively justified as arising from the self-referential character of the problem by [16]:

> the expectational models investors choose affect the price sequence, so that [...] their [...] very choices of models affect their data and so their choices of model.

Arthur et al. [16] refer to the above self-referential structure of the problem as causing it to be ill-defined. Likewise, in the *El Farol* game, [14] gives the classic prototype of a problem with a contrarian structure for which there is no deductive means of arriving at a solution. These problems are well understood to be algorithmically unsolvable [259, 260, 374, 407].

Due to the absence of a unique decision procedure or a deductive means by which to select appropriate forecast models/functions, it is questionable whether rational agents exist at all. The best an agent can do is to find approximations to optimal solutions. Given a limited amount of time, some algorithms will be able to find better approximations than others. Given the same algorithm, the faster a computer is the more potential that it can find better solutions. Therefore, when serious computation is involved, the actual degree of an agent's rationality depends on which algorithms he uses, and what computing power he has access to.

However, computational complexity is not the only reason why the rationality assumption is challenged. Challenges also come from cognitive reasoning [8] which questions the behavior of optimal human beings and proposes a more realistic notion of rationality. This notion is called bounded rationality [369] and describes the property of an agent that behaves in a manner that is nearly as optimal with respect to its goals as its resources will allow. These resources include processing power, algorithm, and time available to the agent. Some of these resources are not always easy to quantify. Therefore, any mathematical model that assumes rationality is subject to the definition of resources available. By using faster computers and better algorithms within a given time limit, it is therefore possible that an agent can become "more rational."

5.3 Computational Difficulties with Efficiency

5.3.1 Information Interpretation

Intellectual incoherency of the rationality concept [304] with the theory of effective computability has also been used as major argument against the notion of market efficiency. According to [103], even if we assume that all actual market analyses would have equal merit (whether supported by human or machine), a new market player could always take advantage of the older players unless the older analyses were not merely the same, but also optimal. The assumption of rationality would therefore imply that market players are equally endowed with intelligence, and that this intelligence is nothing but maximal. It may be thought that by use of the most sophisticated computational analyses, statistical techniques, etc. we may achieve such optimality. But certainly we do not have such an optimal understanding of the markets now. It seems hard to believe that market behavior is the uniquely best understood of all subjects of human science. In fact, most scientists would probably tend to think the opposite. Moreover, the computational difficulties of optimally interpreting information prior to acting in the market are severe in principle. Turing's halting problem tells us that the general case of finding a single computable function amongst the class of all computable functions is undecidable [398]. In the context of market modeling, this would mean that there is no algorithm which will guarantee the discovery of the best market model within finite time. Thus, not even the most sophisticated method of inductive and statistical inference can guarantee to find the best predictor function for any market.

Rather than looking for the very best model of market behavior – requiring of us a general search through all computable functions – we might suppose that a lesser concept of optimality may suffice. That is, by restricting the space of models the computational difficulties will be lessened. For example, we could use linear regression using all available explanatory variables. In that case, there is a fairly quick and simple computer program to find the best linear model of market behavior. At the very least, we should like to select among the explanatory variables, rejecting those which are irrelevant (insignificant) to price predictions. But even for functional forms such as linear variable selection with Akaike's information criterion (AIC) this will be a computationally infeasible, i.e., NP-complete task under the assumption that $P \neq NP$ holds (see [97]). Intuitively, the inclusion of more complicated functional forms as compared to the linear one (e.g., logistic regression) will require more complicated computations. We are therefore left with a choice between grossly inadequate models or computational infeasibility. Thus, we are left either with being satisfied with a suboptimal linear model or with being unable (within feasible time) to find the optimal nonlinear model.

The application of heuristics therefore leads to an imperfect state of nature where rational expectations remain but profit potential arises. As an immediate consequence, agents become opponents which are effectively competing to win. In order to exploit these profit opportunities, agents develop strategies and distinguish among themselves. Thus, it is unlikely that there is any such thing as a typical agent. Furthermore, agents adapt their strategies in the hope of improving performance. Arthur [15] comments on this issue as follows:

> Economic agents, be they banks, consumers, firms or investors, continually adjust their market moves, buying decisions, prices, and forecasts to the situation these moves or decisions or prices or forecasts together create. But unlike ions in a spin glass, which always react in a simple way to their local magnetic field, economic "elements" – human agents – react with strategy and foresight by considering outcomes that might result as a consequence of behavior they might undertake. This adds a layer of complication to economics not experienced in the natural sciences.

Apart from the inherent, and mostly severe, computational difficulties with the discovery of optimal models (and the verification of optimality), there are also technological difficulties standing in the way of the optimal use of optimal models. The reason is that any computation employed by the model will be constrained by the speed and memory constraints imposed by the current level of technology. Therefore, we would not even be able technically to identify the "very best" model of market behavior. Since our analytic techniques are suboptimal and its use is constrained by the sophistication and power of the technology available to us, we can anticipate that the introduction of new generations of computers and other technology will continue to generate new arbitrage opportunities. Statistics not having been "solved," we can also anticipate that the introduction of new techniques and the refinement of existing techniques, of statistical analysis, machine learning, etc.

will supply new means to take advantage of relevant information as it arises. This suggests that arbitrage opportunities exist today and will continue to exist tomorrow. None of which means that the arbitrage opportunities are constant: as old ones are utilized they diminish, while new ones arise.

5.3.2 Computational Complexity of Arbitrage

Computational complexity of arbitrage has been studied for various frictional markets [95], including frictional markets in securities [96] and also foreign exchange markets [67]. A frictional foreign exchange market is characterized by bid-offer spreads, bound constraints, and integrality constraints. Bid-offer spreads reflect the existence of transaction costs which may differ among traders. Bound and integrality constraints refer to limited fixed integer traded amounts in multiples of hundreds, thousands, or millions. For such markets, [67] first demonstrated that finding arbitrage opportunities is an undecideable problem:

Theorem 5.2. *There exists a fixed amount of information $\varepsilon > 0$ available to participating agents such that approximating the optimization version of the arbitrage problem within a factor of n^ε is NP-hard.*

It is only in two special cases that the maximum value of arbitrage revenue can be computed in polynomial time.

Theorem 5.3. *There are polynomial time algorithms for the optimization version of the arbitrage problem if the number of currencies is a constant or the exchange graph is star-shaped (i.e., there exist only exchanges between one "center" currency and the other currencies in the market).*

The results have important consequences towards the fundamental arbitrage-free assumption which is implied in many theories of finance. The efficient markets hypothesis, for instance, is based on the belief that if an arbitrage opportunity ever existed, it would disappear in an arbitrarily short period of time. However, if generating an arbitrage opportunity is a computationally hard problem, then this assumption may not hold in practice. In order for an agent to take advantage of an arbitrage opportunity, he needs to be equipped with all of the information about an exchange in order to generate arbitraging bids that are guaranteed to be in an optimal allocation. However, agents are often constrained by time and computer power. The computational complexity of detecting arbitrage in the foreign exchange markets therefore presents one important reason why arbitrageurs often have to use heuristics to support economic decisions.

Chapter 6
Conclusions

It is difficult for any model to describe adequately, and with a firm empirical basis, all features of modern economies that are relevant to determining exchange rate movements. This reflects in part the difficulty of modeling international financial markets and capital flows. Economists have developed a variety of methods to estimate equilibrium exchange rates [271]. The methods differ considerably in their construction and in their estimations of equilibrium values. In some sense, comparing the models is similar to comparing "apples and oranges" because they can radically differ in structure and can even use different measures of the real effective exchange rate. Often, they are attempting to measure entirely different kinds of equilibrium. That does not mean the models do not provide useful information. To the contrary, they provide valuable insights, but one must recognize that they are limited by the use of somewhat simplified structures, which are often necessary if they are to have a reasonable empirical underpinning.

Complexity matters in economics and finance and in the special case of foreign exchange markets. It prevents easy solutions, efficient algorithms to solve the problems and often even efficient algorithms to approximate optimal solutions. But nevertheless, in real life, decisions have to be made. For instance, despite the complexity, traders in financial firms need to take bets on whether the exchange rate is going to rise or fall within a certain time interval. In addition, risk managers of nonfinancial firms need to decide on when and how to hedge their foreign exchange exposures. The computational complexity of a given problem is therefore an argument that helps agents to choose the right tools to support their decisions. By considering above results on the general computational impossibility of detecting foreign exchange market inefficiencies, we know that it is suitable to use heuristics, which are able to handle problems of high dimensionality such as exchange rate forecasting or approximating the optimal solution to decision problems such as exchange rate hedging.

C. Ullrich, *Forecasting and Hedging in the Foreign Exchange Markets,* Lecture Notes
in Economics and Mathematical Systems 623, DOI: 10.1007/978-3-642-00495-7_6,
© Springer-Verlag Berlin Heidelberg 2009

Part III
Exchange Rate Forecasting with Support Vector Machines

"My experience is that exchange markets have become so efficient that virtually all relevant information is embedded almost instantaneously in exchange rates to the point that anticipating movements in major currencies is rarely possible. [...]. To my knowledge, no model projecting directional movements in exchange rates is significantly superior to tossing a coin." (Alan Greenspan)

Chapter 7
Introduction

Over the past several decades, researchers and practitioners have used various fore-casting methods to study foreign exchange market time series events, thus implicitly challenging the concepts of informational and speculative efficiency. These fore-casting methods largely stemmed from the fields of financial econometrics and ma-chine learning. For example, the 1960s saw the development of a number of large macroeconometric models purporting to describe the economy using hundreds of macroeconomic variables and equations. It was found that although complicated linear models can track the data very well over the historical period, they often per-form poorly for out-of-sample forecasting [287]. This has often been interpreted that the explanatory power of exchange rate models is extremely poor. Nelson [310] dis-covered that univariate autoregressive moving average (ARMA) models with small values for p and q produce more robust results than the big models. Box [58] de-veloped the autoregressive integrated moving average (ARIMA) methodology for forecasting time series events. The basic idea of ARIMA modeling approaches is the assumption of linearity among the variables. However, there are many time series events for which the assumption of linearity may not hold. Clearly, ARIMA mod-els cannot be effectively used to capture and explain nonlinear relationships. When ARIMA models are applied to processes that are nonlinear, forecasting errors often increase greatly as the forecasting horizon becomes longer. To improve forecasting nonlinear time series events, researchers have developed alternative modeling ap-proaches. These include nonlinear regression models, the bilinear model [171], the threshold autoregressive model [392], and the autoregressive heteroskedastic model by [115]. Although these methods have shown improvement over linear models for some specific cases, they tend to be application specific, lack generality, and are of-ten harder to implement [424].

An alternative strategy is for the computer to attempt to learn the input/output functionality from examples, which is generally referred to as supervised learning, a subdiscipline of the machine learning field of research. Machine learning mod-els are rooted in artificial intelligence which distinguishes itself from economic and econometric theory by making statistical inferences without any a priori assump-tions about the data. Intelligent systems are therefore designed to automatically

C. Ullrich, *Forecasting and Hedging in the Foreign Exchange Markets,* Lecture Notes
in Economics and Mathematical Systems 623, DOI: 10.1007/978-3-642-00495-7_7,
© Springer-Verlag Berlin Heidelberg 2009

detect patterns in data, despite complex, nonlinear behavior.[1] If the patterns detected are significant, a system can expect to make predictions about new data coming from the same source. An intelligent system can thus acquire generalization power by learning something about the source generating the data. Intelligent systems are therefore data driven learning methodologies that seek to approximate optimal solutions of problems of high dimensionality. Since, in contrast, theory driven approaches give rise to precise specifications of the required algorithms for solving simplified models the search for patterns replaces the search for reasons.

Artificial neural networks (ANN) are general-purpose self-learning systems that grew out of the cognitive and brain science disciplines for approximating the way information is being processed [196]. Instead of analyzing the vertical relationship between underlying cause and its derived effect, ANN learning models focus on how effects reproduce themselves horizontally and what this reveals about their inherent dynamics. Hence, instead of a linear cause–effect relation, ANN models are able to exploit nonlinear data relationships by moving the effect in a constant feedback loop. During the last decade, the application of ANNs as supervised learning methods has exploded in a variety of areas [191, 379, 425]. Within the realm of financial forecasting, ANNs have been used to develop prediction algorithms for financial asset prices, such as technical trading rules for stocks and commodities [130, 221, 223, 365, 377, 416, 421]. The effectiveness of ANNs and their performance in comparison to traditional forecasting methods has also been a subject of many studies [91, 426]. ANNs have proven to be comprehensive and powerful for modeling nonlinear dependencies in financial markets [336], notably for exchange rates [48, 107, 234, 306]. However, ANN models have been criticized because of their black-box nature, excessive training times, danger of overfitting, and the large number of parameters required for training. As a result, deciding on the appropriate network involves much trial and error. These shortcomings paired with the logic that complex real-world problems might require more elaborate solutions led to the idea of combining ANNs with other technologies to hybrid and modular solutions [9]. For a survey of the application of ANN to forecasting problems in general see [424, 426].

Support vector machines (SVM) [55, 405] are a new kind of supervised learning system which, based on the laws of statistical learning theory [405], maps the input dataset via kernel into a high dimensional feature space in order to enable linear data classification and regression. SVM has proven to be a principled and very powerful method that in the few years since its introduction has already outperformed many other systems in a variety of applications, such as text categorization [207], image processing [328, 333], hand-written digit recognition [247], and bioinformatic problems, for example, protein homology detection [203] and gene expression [64]. Subsequent applications in time series prediction [301] further indicated the potential that SVMs have with respect to the economic and financial audience. In the special case of predicting Australian foreign exchange rates, [219] showed that moving average-trained SVMs have advantages over an ANN based model which

---

[1] By patterns we understand any relations, regularities, or structure inherent in a given dataset.

was shown to have advantages over ARIMA models [217]. Kamruzzaman et al. [218] had a closer look at SVM regression and investigated how they perform with different standard kernel functions. It was found that Gaussian radial basis and polynomial kernels appeared to be a better choice in forecasting the Australian foreign exchange market than linear or spline kernels. However, although Gaussian kernels are adequate measures of similarity when the representation dimension of the space remains small, they fail to reach their goal in high dimensional spaces [131].

This shortcoming has been addressed by [400–402]. They examined the suitability of SVMs, equipped with p-Gaussian, as well as several standard kernel functions and trained with exogenous financial market data in order to forecast EUR exchange rate ups and downs. This research will be presented in detail in the following. The task is to examine the potential of SVM models to overcome foreign exchange market efficiency in its weak-form. We therefore analyze the ability of SVMs to correctly classify daily EUR exchange rate return data. If an SVM model is able to outperform the nave prediction by exploiting nonlinear patterns in daily EUR/GBP, EUR/JPY, and EUR/USD exchange rate returns, market efficiency in its weak form can be rejected. Indeed, it is more useful for traders and risk managers to forecast exchange rate fluctuations rather than their levels. To predict that the level of the EUR/USD, for instance, is close to the level today is trivial. On the contrary, to determine if the market will rise or fall is much more interesting. Since SVM performance depends to the most extent on choosing the right kernel, we empirically verify the use of customized p-Gaussians by comparing them with a range of standard kernels.

The chapter is organized as follows: in the next section, we conduct statistical analyses of EUR/GBP, EUR/JPY, and EUR/USD time series. Chapter 9 outlines the principles of support vector classification including a description of the binary classification problem which is the theoretic problem under consideration, and the idea of learning in feature space. Chapter 10 describes the details of the empirical study, such as the procedure for obtaining an explanatory input dataset (description model) as well as the particular SVM model and kernel functions used (forecasting model). Chapter 10 also provides the benchmarks and metrics used for model evaluation along with the results.

Chapter 8
Statistical Analysis of Daily Exchange Rate Data

8.1 Time Series Predictability

A time series $\{y_t\}$ is a discrete time continuous state process where the variable y is identified by the value that it takes at time t denoted y_t. Time is taken at equally spaced intervals from $-\infty$ to $+\infty$ and the finite sample size T of data on y is for $t = 1, 2, \ldots, T$. Time series $\{y_t\}$ may emerge from deterministic and/or stochastic influences. For example, a time trend $y_t = t$ is a very simple deterministic time series. If $\{y_t\}$ is generated by a deterministic linear process, it has high predictability, and its future values can be forecasted very well from the past values. A basic stochastic time series is *white noise*, $y_t = \varepsilon_t$, where ε_t is an independent and identically distributed (i.i.d.) variable with mean 0 and variance σ^2 for all t, written $\varepsilon_t \sim i.i.d.(0, \sigma^2)$. A special case is *Gaussian white noise*, where the ε_t are independent and normally distributed variables with mean 0 and variance σ^2 for all t, written $\varepsilon_t \sim NID(0, \sigma^2)$. A time series generated by a stochastic process has low predictability, and its past values provide only a statistical characterization of the future values. Predictability of a time series can therefore be considered as the signal strength of the deterministic component of the time series to the whole time series. Usually, a given time series is not simply deterministic or stochastic, but rather some combination of both:

$$y_t = \alpha + \beta t + \varepsilon_t \tag{8.1}$$

8.2 Empirical Analysis

The purpose of this section is to examine the statistical properties of daily EUR/GBP, EUR/JPY, and EUR/USD exchange rate data from 1 January 1997 to 31 August 2003. This is done for two reasons. First, according to above, time series analysis gives an understanding on the degree of randomness inhibited in the chosen time interval. The strategy is to build econometric models in order to extract

C. Ullrich, *Forecasting and Hedging in the Foreign Exchange Markets,* Lecture Notes in Economics and Mathematical Systems 623, DOI: 10.1007/978-3-642-00495-7_8,
© Springer-Verlag Berlin Heidelberg 2009

statistical dependencies within a time series that may be based on linear and/or
nonlinear relationships. If such dependencies are significant, then the time series
is not totally random since it contains deterministic components. These may be
important indicators for the predictability of $\{y_t\}$. Second, analysis of $\{y_t\}$ is useful
for building empirical time series models that will serve as benchmarks for our
SVM models. The advantage of our data-driven approach is that it allows variables
to speak for themselves, without the confines of economic or financial theories.

The investigation is based on publicly available London daily closing prices as
obtained from http://www.oanda.com/. When examining daily data, closing prices
are more relevant than opening prices since they represent the matching of supply
and demand at the end of the trading day. The series for the period from 1997 to
1998 were constructed by using the fixed EUR/DEM conversion rate agreed in
1998 (1 EUR = 1.95583 DEM), combined with the GBP/DEM, JPY/DEM, and
USD/DEM daily market rates. It is also important to note that we do not include the
period from 1 September 2003 to 31 December 2004 in our analysis since it will be
needed for out-of-sample forecasting and is not known beforehand.

Plots of the three time series are shown in Figs. 8.1–8.3 which demonstrate the
devaluation of the EUR since its introduction in 1998 until the end of 2000.

The devaluation had been followed by a rapid upward trend against the three
foreign currencies reaching to the end of the specified time window. A comparison of the magnitude of these upward and downward movements is visualized by
Fig. 8.4, which shows that among the three exchange rates EUR/JPY has been the
most volatile one, and EUR/GBP the least volatile one. The vertical bar drawn at
29 August 2003 divides the series into an insample period (left of the bar) and and
out-of-sample period (right of the bar).

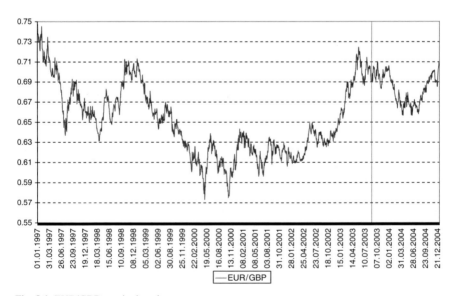

Fig. 8.1 EUR/GBP nominal exchange rate

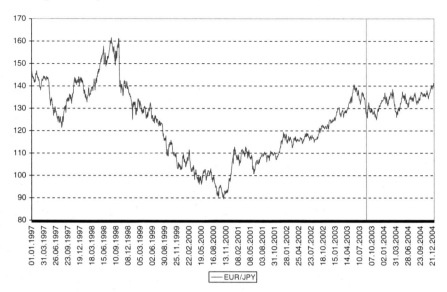

Fig. 8.2 EUR/JPY nominal exchange rate

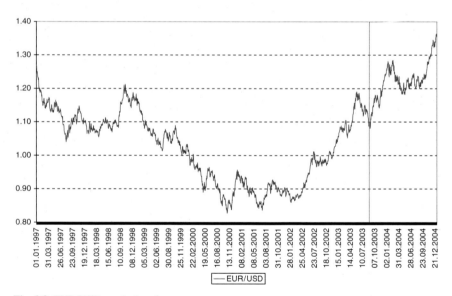

Fig. 8.3 EUR/USD nominal exchange rate

For means of comparison, all nominal exchange rates were scaled to basis 100 on 1 January 1997. The paths were then calculated on a continuously compounded basis by sequentially adding log first differences.

Fig. 8.4 Base currency EUR in terms of foreign currencies GBP, JPY, and USD

Every time series was examined in a hypothesis testing framework according to the following criteria:

1. Stationarity
2. Normality
3. Linearity
4. Heteroskedasticity
5. Nonlinearity

In a hypothesis testing framework, there are always two hypotheses that go together, known as the null hypothesis (H_0) and the alternative hypothesis (H_1). The null hypothesis is the statement or the statistical hypothesis that is actually being tested. The alternative hypothesis represents the remaining outcomes of interest. Hypothesis tests were conducted via a test of significance approach which centers on a statistical comparison of the estimated value of the coefficient and its value under the null hypothesis. Generally, if the estimated value is a long way away from the hypothesized value, the null hypothesis is likely to be rejected. Otherwise, if the value under the null hypothesis and the estimated value are close to one another, the null hypothesis is less likely to be rejected. Note that it is incorrect to state that if the null hypothesis is not rejected it is accepted since it is impossible to know whether the null is actually true or not.

8.2.1 Stationarity

As a first step we ensured that the time series data we work with are stationary, which is an important property to apply for statistical inference procedures. A series is said to be strictly stationary, if the mean, the variance, and the covariance of the underlying series remain constant over time ([180], p. 45):

$$E(y_t) = \mu = const.$$
$$Var(y_t) = E\left((y_t - \mu)^2\right) = \sigma^2 = const. \quad (8.2)$$
$$Cov(y_t, y_{t+\tau}) = E\left((y_t - \mu)(y_{t+\tau} - \mu)\right) = \lambda = const.$$

However, strict stationarity as represented in (8.2) is generally not fulfilled in financial and economic time series. For this reason, econometric theory shows a greater interest in whether a time series is (weakly or covariance) stationary, which is the case if the mean and variance are independent of time t, and if the series' covariance only depends on the time difference τ, but not on t itself:

$$Cov(y_t, y_{t+\tau}) = E\left((y_t - \mu)(y_{t+\tau} - \mu)\right) = \lambda_\tau \quad (8.3)$$

Statistical tests of the null hypothesis that a time series is nonstationary against the alternative that it is stationary are called unit root tests. The name derives from the fact that a stochastic process is nonstationary if the characteristic polynomial has a root that does not lie inside the unit circle. The unit root tests applied are the augmented Dickey–Fuller (ADF) test [98], and the Philipps–Perron (PP) test [324]. The Kwiatkowski–Philipps–Schmidt–Shin (KPSS) test [236] which is additionally applied differs from the above two tests in that the underlying time series is assumed to be stationary under the null hypothesis.

Table 8.1 shows that the results of the three tests are consistent for the series of exchange rate levels y_t. In the case of the unit root tests, we cannot reject the null hypothesis that daily price data are generated by a nonstationary stochastic process. Otherwise, in the case of the KPSS test, we reject the null hypothesis that daily price data are generated by a stationary stochastic process at the 1% significance level which matches the result of the unit root tests.

A time series that is nonstationary, but can be stationarized by d-fold differentiation is called *integrated* of order d, i.e., $I(d)$ [58]. Accordingly, log first differences $\Delta y_t = \ln(y_t) - \ln(y_{t-1})$ of the price data for the period from 2 January 1997 to 31 December 2003 were taken, as visualized by Figs. 8.5–8.7, the same tests were

Table 8.1 Testing for (non-) stationarity

Null hypothesis	Test	Input	Output		
			EUR/GBP	EUR/JPY	EUR/USD
Nonstationarity	ADF	$y_t, \Delta y_t$	$-2.28, -43.98^{***}$	$-1.31, -40.48^{***}$	$-1.34, -44.70^{***}$
	PP	$y_t, \Delta y_t$	$-2.10, -44.14^{***}$	$-1.33, -40.47^{***}$	$-1.41, -44.64^{***}$
Stationarity	KPSS	$y_t, \Delta y_t$	$0.79^{***}, 0.03$	$0.90^{***}, 0.07$	$0.81^{***}, 0.07$

The columns from left to right denote the null hypothesis, name of the test applied, time series input, and the test statistic output per currency pair. *, **, and *** indicate rejection of the null hypothesis at the 10%, 5%, and 1% significance level

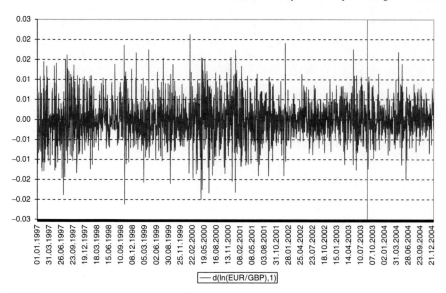

Fig. 8.5 Log first differences of daily EUR/GBP nominal exchange rate data

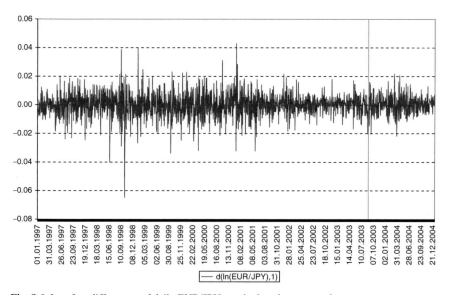

Fig. 8.6 Log first differences of daily EUR/JPY nominal exchange rate data

conducted subsequently. Again, the vertical bar drawn at 29 August 2003 divides the series into in-sample period (left of the bar) and out-of-sample period (right of the bar).

Test statistics suggest that now all three exchange rate series are strongly difference-stationary, i.e., integrated of order one (I(1)).

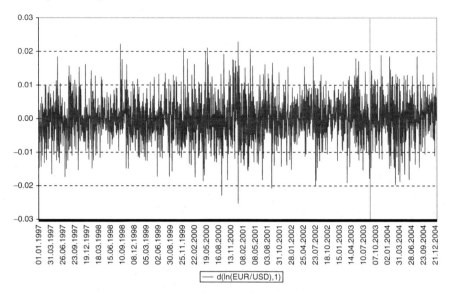

Fig. 8.7 Log first differences of daily EUR/USD nominal exchange rate data

8.2.2 Normal Distribution

In a second step, the empirical probability distributions of Δy_t were tested for departures from the normal distribution. The normality assumption $\Delta y_t \sim N(0, \sigma^2)$ is required in order to conduct hypothesis tests about the model parameters. One of the most commonly applied tests for normality is the Jarque–Bera (JB) test [206]. JB uses the property of a normally distributed random variable which characterizes the entire distribution by its first two moments, i.e., the mean and the variance. The standardized third and fourth moments of a distribution are known as its skewness (SK) and kurtosis (KU). The coefficient of SK measures the extent to which a distribution is not symmetric about its mean value and is expressed as

$$SK = \frac{E[y^3]}{(\sigma^2)^{3/2}} \tag{8.4}$$

A normal distribution is defined to have $SK = 0$. KU measures how fat the tails of the distribution are and is expressed as

$$KU = \frac{E[y^4]}{(\sigma^2)^2} \tag{8.5}$$

A normal distribution is defined to have a coefficient of kurtosis of $KU = 3$. Instead of KU, it is also common to use the coefficient of excess kurtosis (EK), given by $EK = KU - 3$, which must be zero for a normal distribution. JB tests

Table 8.2 Testing for normality

Null hypothesis	Test	Input	Output		
			EUR/GBP	EUR/JPY	EUR/USD
–	SK	Δy_t	0.09	−0.34	0.06
–	KU	Δy_t	3.74	6.85	3.62
Normal Distribution	JB	Δy_t	42.11***	1106.89***	29.18***

The columns from left to right denote the null hypothesis, name of the test applied, time series input, and the test statistic output per currency pair. *, **, and *** indicate rejection of the null hypothesis at the 10%, 5%, and 1% significance level

whether the coefficient of *SK* and the coefficient of *EK* are jointly zero by the test statistic

$$JB = T\left[\frac{SK^2}{6} + \frac{(KU-3)^2}{24}\right] \qquad (8.6)$$

where T is the sample size. The test statistic asymptotically follows a $\chi^2(2)$ under the null hypothesis that the series is symmetric and mesokurtic.

Table 8.2 summarizes the results of our tests for normality. We observe that EUR/GBP and EUR/USD are slightly skewed to the right (positive signs) whereas EUR/JPY is skewed to the left (negative sign). We further observe that all three series have positive excess kurtosis indicating that there is more weight in both tails of the distribution than in the normal distribution. Probability distributions exhibiting this phenomenon are said to be leptokurtic or fat-tailed. Among the three series, EUR/JPY exhibits the most leptokurtic behavior whereas EUR/GBP and EUR/USD show weaker signs of fat tails. This is confirmed by the JB test statistics which reject the null hypothesis of normal data at high levels. The empirical observation of leptokurtic exchange rate returns has been mentioned in countless studies (see, for instance, [375], p. 148).

8.2.3 Linearity

A major objective when analyzing stationary time series is to detect linear dependencies among the data through identifying an appropriate linear model. Univariate time series models can only be explained by their own lagged values, i.e., by autoregressive (AR) terms as explanatory variables in their representation. The autoregressive model of order 1, the AR(1) model is

$$y_t = c + \alpha y_{t-1} + \varepsilon_t \qquad (8.7)$$

where $\varepsilon_t \sim i.i.d.(0, \sigma^2)$. The constant term c models a trend in the series either upwards ($c > 0$) or downwards ($c < 0$). The lag coefficient α determines the stability of the process. If $|\alpha| > 1$ the time series will explode, that is, $y_t \to \pm\infty$ as $t \to \infty$.

The special case $|\alpha| = 1$ gives the random walk model and it is only when $|\alpha| < 1$ that the process defined by (8.7) will be stationary.

Furthermore, if the underlying process is stochastic and stationary, the errors can be linear combinations of white noise at different lags, so the moving average (MA) part of the model refers to the structure of the error term ε_t. The first-order moving average model MA(1) is

$$y_t = c + \varepsilon_t + \beta \varepsilon_{t-1} \tag{8.8}$$

where $\varepsilon_t \sim i.i.d.(0, \sigma^2)$. This model is a stationary representation for any values of c or b, since $E(y_t) = c$. The most general model for a stationary process is an integrated autoregressive moving average model ARIMA(p, q, r)

$$y_t = c + \alpha_1 y_{t-1} + \alpha_2 y_{t-2} + \ldots + \alpha_p y_{t-p} + \varepsilon_t + \beta_1 \varepsilon_{t-1} + \beta_q \varepsilon_{t-q} \tag{8.9}$$

with p autoregressive terms, q moving average terms, integration order r, with $r = 1$ in our case, and $\varepsilon_t \sim i.i.d.(0, \sigma^2)$. ARIMA($p, q, 1$) models are also simply denoted as ARMA(p, q) models. The parameters p and q are commonly estimated by visually inspecting the autocorrelation function (ACF) and partial autocorrelation function (PACF) for MA models and low-order AR models [180]. ACF and PACF functions characterize the pattern of temporal, linear dependence that is existent in the series. Since independent variables are always uncorrelated, testing for linear independency is equivalent to testing for zero autocorrelation. We first calculated Durbin–Watson (DW) test statistics [108] for Δy_t. DW is a test for first order autocorrelation, i.e., it only tests for a relationship between an error and its immediately previous value. The test statistic is

$$DW \approx 2(1 - \hat{\rho}) \tag{8.10}$$

with $\hat{\rho} \in [-1, 1]$ denoting the correlation between Δy_t and Δy_{t-1}. Thus if DW $= 2$ $\hat{\rho} = 0$ there is no autocorrelation in the residuals and the null hypothesis would not be rejected. Otherwise if DW $= 0$ ($\hat{\rho} = 1$) there is perfect positive autocorrelation in the residuals and, if DW $= 4$ ($\hat{\rho} = -1$) there is perfect negative autocorrelation in residuals. The DW test does not follow a standard statistical distribution but is divided up into regions including an upper and lower critical value.

Second, Ljung–Box (LB) Q-statistics [266]

$$Q^* = T(T+2) \sum_{k=1}^{m} \frac{\hat{\tau}_k^2}{T-k} \sim \chi^2 \tag{8.11}$$

were calculated to test the joint hypothesis that all m of the τ_k correlation coefficients are simultaneously equal to zero.

It is interesting to see from Table 8.3 that for the EUR/JPY series no significant linear relationships among the data can be identified and thus, one would not necessarily need to estimate a linear model.

However, we decided at this point to obtain clearer statistical results by also removing some insignificant linearities through

$$\Delta y_t = \underset{(1.20)}{0.0287} \Delta y_{t-1} + \varepsilon_t \tag{8.12}$$

Table 8.3 Testing for linear dependencies

Null hypothesis	Test	Input	Output		
			EUR/GBP	EUR/JPY	EUR/USD
No AC	LB	Δy_t, ARMA- Residuals of Δy_t	$k = 1$: 4.95**, 0.00	$k = 1$: 1.42, 0.00	$k = 1$: 8.36***, 0.03
			$k = 2$: 4.98*, 0.13	$k = 2$: 1.44, 0.02	$k = 2$: 10.49***, 0.07
			$k = 3$: 10.43**, 0.14	$k = 3$: 2.00, 0.56	$k = 3$: 11.85***, 0.07
			$k = 4$: 10.60**, 0.15	$k = 4$: 2.41, 0.97	$k = 4$: 13.07**, 0.34
			$k = 5$: 10.61**, 0.17	$k = 5$: 3.02, 1.67	$k = 5$: 13.18**, 0.54
			$k = 6$: 10.65*, 0.18	$k = 6$: 4.38, 3.13	$k = 6$: 13.21**, 0.55
			$k = 7$: 10.70, 0.23	$k = 7$: 4.48, 3.24	$k = 7$: 14.10**, 1.66
			$k = 8$: 12.38, 2.23	$k = 8$: 5.24, 4.19	$k = 8$: 15.74**, 2.99
			$k = 9$: 13.39, 3.20	$k = 9$: 5.52, 4.29	$k = 9$: 16.83*, 3.06
			$k = 10$: 14.19, 3.74	$k = 10$: 6.59, 5.27	$k = 10$: 16.79*, 4.44
			$k = 15$: 16.44, 5.58	$k = 15$: 9.67, 8.27	$k = 15$: 21.03, 7.55
			$k = 20$: 22.49, 11.46	$k = 20$: 15.70, 14.71	$k = 20$: 23.88, 11.78
			$k = 24$: 32.93, 19.77	$k = 24$: 15.84, 14.85	$k = 24$: 26.82, 14.51
	BG	ARMA- Residuals of Δy_t	0.8617	0.6836	0.8975
	DW	Δy_t, ARMA- Residuals of Δy_t	1.9939, 1.9988	1.9995, 1.9995	1.9928, 2.0021

The columns from left to right denote the null hypothesis, name of the test applied, time series input, and the test statistic output per currency pair. *, **, and *** indicate rejection of the null hypothesis at the 10%, 5%, and 1% significance level. k denotes the number of lags included in the test regression

For the other two series, linear dependencies are obviously existent. To remove them, it was sufficient to specify the following simple linear models

$$\Delta y_t = -0.0526\Delta y_{t-1} - 0.0562\,\Delta y_{t-3} + \varepsilon_t \qquad (8.13)$$
$$\quad\;\;(-2.20) \qquad\quad (-2.35)$$

for the EUR/GBP series and

$$\Delta y_t = -0.5959\Delta y_{t-1} + 0.5323\varepsilon_{t-1} + \varepsilon_t \qquad (8.14)$$
$$\quad\;\;(-3.01) \qquad\quad (2.55)$$

for the EUR/USD series. T-statistics are given in brackets below the parameter estimates. Each model's residuals series $\{\varepsilon_t\}$ was retested according to LB and the Breusch–Godfrey Lagrange Multiplier (LM) test which, like LB, is a more general test for autocorrelation up to the rth order. The model for the errors under this test is

$$\varepsilon_t = \beta_1\varepsilon_{t-1} + \beta_2\varepsilon_{t-2} + \ldots + \beta_r\varepsilon_{t-r} + v_t \qquad (8.15)$$

and the null hypothesis is that the current error is not related to any of its r previous values, i.e., $\beta_1 = \beta_2 = \ldots = \beta_r = 0$. This kind of test is known as a test on omitted variables. The test statistic of the Breusch–Godfrey LM test is given by

$$(T - r)R^2 \sim \chi_r^2 \tag{8.16}$$

The test statistics indicate that serial dependencies have now disappeared at any lag. Hence, the models are statistically adequate. However, although linear independency can be inferred for all three series, nonlinear dependencies might still exist.

8.2.4 Heteroskedasticity

In a next step, we investigated the origin of nonnormal behavior by focusing on the phenomenon of heteroskedastic processes [115]. The classical regression model assumes that the error process ε_t in the model is homoskedastic. In other words, ε_t is assumed to have a constant variance $\text{Var}(\varepsilon_t) = \sigma^2$. Heteroskedasticity is motivated by the observation that in many financial time series the variance of et appears to be related to the variance of recent residuals, i.e., $\text{Var}_t(\varepsilon_t) = \sigma_t^2$. This phenomenon which is referred to as *volatility clustering* [45, 273], indicates that returns Δy_t are not in fact independent but rather, they follow some sort of nonlinear dynamic process. In order to detect these second-moment dependencies, we tested according to [286]. We first calculated the autocorrelations of the squared residuals and computed the LB Q-statistics for the corresponding lags. If there is no autoregressive conditional heteroskedasticity (ARCH) in the residuals, autocorrelations and partial autocorrelations should be zero at all lags and the Q-statistics should not be significant. In order to find out whether these results can be confirmed by a different test, we apply the ARCH LM test which tests the null hypothesis that there is no ARCH up to order q in the residuals. The test statistic is computed from the auxiliary regression

$$\varepsilon_t^2 = \beta_0 + \beta_1 \varepsilon_{t-1}^2 + \beta_2 \varepsilon_{t-2}^2 + \ldots + \beta_q \varepsilon_{t-q}^2 + v_t \tag{8.17}$$

where ε is the residual. The F-statistic is used as an omitted variable test for the joint significance of all lagged squared residuals.

According to Table 8.4, LB outputs reject the null hypothesis of "No ARCH" at very high significance levels for all three series of squared ARMA residuals. The ARCH-LM F-statistics confirms the results for the squared ARMA residuals of EUR/GBP and EUR/JPY: the null hypothesis of zero heteroskedasticity is clearly rejected for all selected lags at the 1% level.

The result for EUR/USD is less clear: according to both, Q-statistics and ARCH-LM testing results, the hypothesis of a constant variance can only be rejected at higher lags and with slightly lower confidence. This brings us to an important result, which has also been reported in literature ([99], p. 10): ARCH processes are leptokurtic, or *fat-tailed*, relative to the normal. The slightly weaker test statistics for EUR/USD may perhaps be explained by a skewness and kurtosis that are closer

Table 8.4 Testing for Heteroskedasticity

Null	Test	Input	Output		
			EUR/GBP	EUR/JPY	EUR/USD
No ARCH	LB	(ARMA-, Residuals)^2 of Δy_t, (GARCH-Residuals)^2 of Δy_t	$k=1$: 20.40, 0.06 $k=2$: 27.86, 0.55 $k=3$: 52.93***, 0.67 $k=4$: 56.64***, 0.69 $k=5$: 76.08***, 0.70 $k=6$: 87.65***, 0.78 $k=7$: 111.06***, 2.57 $k=8$: 113.33***, 3.86 $k=9$: 117.60***, 3.88 $k=10$: 124.61***, 4.02 $k=15$: 143.77***, 5.35 $k=20$: 172.06***, 11.12	$k=1$: 82.96, 0.59 $k=2$: 97.67***, 1.18 $k=3$: 98.78***, 1.58 $k=4$: 107.73***, 1.58 $k=5$: 117.94***, 1.62 $k=6$: 131.67***, 6.05 $k=7$: 132.71***, 6.21 $k=8$: 137.94***, 6.86 $k=9$: 141.91***, 6.94 $k=10$: 142.09***, 8.15 $k=15$: 151.57***, 10.13 $k=20$: 209.19***, 12.42	$k=1$: 0.21, 0.01 $k=2$: 0.35, 0.19 $k=3$: 5.09**, 0.41 $k=4$: 5.41*, 0.70 $k=5$: 11.42**, 1.34 $k=6$: 15.64***, 1.73 $k=7$: 17.12***, 1.77 $k=8$: 17.84***, 1.77 $k=9$: 20.37***, 1.96 $k=10$: 23.13***, 1.99 $k=15$: 38.50***, 5.76 $k=20$: 46.12***, 11.42
	ARCH LM	ARMA-Residuals of Δy_t, (GARCH-Residuals)^2 of Δy_t	$k=1$: 20.59***, 0.06 $k=4$: 11.91***, 0.18 $k=8$: 10.15***, 0.48 $k=12$: 7.34***, 0.42	$k=1$: 86.81***, 0.59 $k=4$: 24.61***, 0.39 $k=8$: 14.02***, 0.85 $k=12$: 9.74***, 0.73	$k=1$: 0.21, 0.01 $k=4$: 1.33, 0.17 $k=8$: 2.03**, 0.21 $k=12$: 1.74*, 0.32

The columns from left to right denote the null hypothesis, name of the test applied, time series input, and the test statistic output per currency pair. *, **, and *** indicate rejection of the null hypothesis at the 10%, 5%, and 1% significance level. k denotes the number of lags included in the test regression

to the ones of a normal distribution. It is possible to remove heteroskedastic effects within the time series by estimating ARCH models, a family of models that was introduced by [115] and generalized as GARCH (Generalized ARCH) by [49]. A GARCH model consists of two equations. The first is the conditional mean equation which has already been estimated in (8.12)–(8.14). The second equation is the conditional variance equation whose form determines the type of GARCH model.

Academic literature has proposed many different types of GARCH models (see [50,51,318] for good surveys). The generic GARCH(1,1) model with one error term and one autoregressive term is

$$\sigma_t^2 = (1-\alpha-\beta)\sigma^2 + \alpha\varepsilon_{t-1}^2 + \beta\sigma_{t-1}^2 \qquad (8.18)$$
$$= \sigma^2 + \alpha(\varepsilon_{t-1}^2 - \sigma^2) + \beta(\sigma_{t-1}^2 - \sigma^2)$$

where

$$\sigma^2 = \frac{\omega}{1-\alpha-\beta} \qquad (8.19)$$

with ω denoting the GARCH constant, $\alpha \geq 0$ denoting the GARCH error coefficient, and $\beta \geq 0$ denoting the GARCH lag coefficient. An ordinary ARCH model is a special case of a GARCH specification in which there are no lagged forecast variances in the conditional variance equation. The GARCH(1,1) process is the most prominent specification for GARCH volatility models, being relatively easy to estimate and generally having robust coefficients that are interpreted naturally in terms of longterm volatilities and short-run dynamics. Many financial markets, including stock and bond markets, have been successfully characterized by such models. Since there is a convergence in term structure forecasts to the long-term average volatility level, the time series of any GARCH volatility forecast will be stationary. However, currencies tend to have volatilities that are not as mean-reverting as the ones of other types of financial assets [151] or that may not mean-revert at all. In this case the usual stationary GARCH models will not apply. Thus it might be more useful in the currency markets to use a component GARCH model in order to regain the convergence in GARCH term structures, by allowing for a time-varying long-term volatility [117, 119, 120]. In components GARCH, σ^2 is replaced by a time-varying *permanent* component given by

$$q_t = \omega + \rho(q_{t-1} - \omega) + \zeta(\varepsilon_{t-1}^2 - \sigma_{t-1}^2) \tag{8.20}$$

Therefore the conditional variance equation in the components GARCH model is

$$\sigma_t^2 = q_t + \alpha(\varepsilon_{t-1}^2 - q_{t-1}) + \beta(\sigma_{t-1}^2 - q_{t-1}) \tag{8.21}$$

Equations (8.20) and (8.21) together define the components model. If $\rho = 1$, the permanent component to which long-term volatility forecasts mean-revert is just a random walk. While the components model has an attractive specification for currency markets, parameter estimates may lack robustness and therefore specification has to pass diagnostic tests. We estimated the component GARCH model for EUR/GBP returns as

$$\Delta y_t = \underset{(-1.56)}{-0.0382\Delta y_{t-1}} \underset{(-1.86)}{-0.0462\Delta y_{t-3}} + \varepsilon_t$$

$$q_t = \underset{(7.31)}{3.05E-05} + \underset{(137.39)}{0.9825}(q_{t-1} - \underset{(7.31)}{3.05E-05}) \tag{8.22}$$

$$+ \underset{(5.59)}{0.0513}(\varepsilon_{t-1}^2 - \sigma_{t-1}^2)$$

$$\sigma_t^2 = q_t + \underset{(4.67)}{0.0568}(\varepsilon_{t-1}^2 - q_{t-1}) - \underset{(-29.01)}{0.9092}(\sigma_{t-1}^2 - q_{t-1})$$

For EUR/USD returns the following model was estimated:

$$\Delta y_t = \underset{(-3.24)}{-0.6253\Delta y_t} - 1 + \underset{(2.79)}{0.5666\varepsilon_{t-1}}$$

$$q_t = \underset{(11.78)}{4.11E-05} + \underset{(161.67)}{0.9869}(q_{t-1} - \underset{(11.78)}{4.11E-05}) \tag{8.23}$$

$$+ \underset{(3.29)}{0.0227}(\varepsilon_{t-1}^2 - \sigma_{t-1}^2)$$

$$\sigma_t^2 = q_t - \underset{(-2.02)}{0.0348}(\varepsilon_{t-1}^2 - q_{t-1}) + \underset{(-1.33)}{0.5296}(\sigma_{t-1}^2 - q_{t-1})$$

The estimates of persistence in the long run component are $\hat{\rho} = 0.9825$ for EUR/GBP and $\hat{\rho} = 0.9869$ for EUR/USD, indicating that in both models the long run component converges very slowly to the steady state. The short run volatility component also appears to be significantly different from zero in both models. LB Q-statistics as well as the F-statistics of the ARCH-LM test in Table 8.4 demonstrate that second-moment dependencies in the squared residuals of the component GARCH models have now disappeared at any lag. However, this could not be achieved for EUR/JPY returns. The reason could be that volatility is higher in a falling EUR/JPY market than it is in a rising market. This observation is known as *leverage effect* (see [120]). In other words volatility in the EUR/JPY market is an asymmetric phenomenon which cannot be captured by a symmetric component GARCH model that is only able to model ordinary volatility clustering. Consequently, for EUR/JPY it is better to specify an asymmetric component GARCH model

$$q_t = \omega + \rho(q_{t-1} - \omega) + \phi(\varepsilon_{t-1}^2 - \sigma_{t-1}^2) + \theta_1 z_{1t}$$

$$\sigma_t^2 = q_t + \alpha(\varepsilon_{t-1}^2 - q_{t-1}) + \gamma(\varepsilon_{t-1}^2 - q_{t-1})d_{t-1} \qquad (8.24)$$

$$+ \beta(\sigma_{t-1}^2 - q_{t-1}) + \theta_2 z_{2t}$$

where $d_t = 1$ if $\varepsilon_t < 0$, and 0 otherwise. In this model positive shocks ($\varepsilon_t > 0$), and negative shocks ($\varepsilon_t < 0$) have different effects on the conditional variance. Positive shocks have an impact of α, while negative shocks have an impact of $(\alpha + \gamma)$. If $\gamma > 0$ we say that a leverage effect exists in that negative shocks increase volatility. If $\gamma \neq 0$, the news impact is asymmetric. The asymmetric component model for EUR/JPY returns was estimated as

$$\Delta y_t = \underset{(1.29)}{0.0314} \Delta y_{t-1} + \varepsilon_t$$

$$q_t = \underset{(5.14)}{6.75E - 05} + \underset{(339.60)}{0.9932}(q_{t-1} - \underset{(5.14)}{6.75E - 05})$$

$$+ \underset{(5.38)}{0.0369}(\varepsilon_{t-1}^2 - \sigma_{t-1}^2) \qquad (8.25)$$

$$\sigma_t^2 = q_t - \underset{(-1.94)}{0.0477}(\varepsilon_{t-1}^2 - q_{t-1}) + \underset{(5.91)}{0.2132}(\varepsilon_{t-1}^2 - q_{t-1})d_{t-1}$$

$$- \underset{(-0.01)}{0.0013}(\sigma_{t-1}^2 - q_{t-1})$$

We observe that in (8.25) the leverage effect term γ is significantly positive and so it appears that there is indeed an asymmetric effect. Note that it is important that we use the quasi-likelihood robust standard errors (z-values in brackets) since the residuals are highly leptokurtic.

8.2.5 Nonlinearity

Finally, we investigated the existence of nonlinear dependencies within the data. Hsieh [197] divides the realm of nonlinear dependencies into two broad categories – additive nonlinear dependence and multiplicative nonlinear dependence. Additive nonlinearity, also known as nonlinearity-in-mean, enters a process through its mean or expected value, so that each element in the sequence can be expressed as the sum of a zero-mean random element and a nonlinear function of past elements. With multiplicative nonlinearity, or nonlinearity-in-variance, each element can be expressed as the product of a zero-mean random element and a nonlinear function of past elements, so that the nonlinearity affects the process through its variance. Lee [248] examined the performance of a range of tests on nonlinear dependencies across a variety of data generating processes. They find that no single test dominates all the others. In the light of this finding, it is advisable to use more than one test. One of the most general and widely used tests for detecting nonlinear dependencies in a time series is the BDS test [60]. The BDS statistic has its origins in the correlation dimension plots of [175], which were developed for studying low-dimensional chaos in time series in physics applications. A chaotic dynamic process is a complex, but deterministic, nonlinear dynamic process. A simple example is the logistic map $y_t = 4y_{t-1}(1 - y_{t-1})$ where $y_t \in (0,1)$. Such a process may look random, but is, at least in theory, potentially perfectly predictable. However, financial time series are more likely to follow nonlinear dynamic processes that are stochastic, rather than ones that are chaotic and deterministic (additive nonlinearities). The BDS statistic was developed to detect the existence of any type of either of these two categories of nonlinear dynamics. Thus if the null hypothesis of i.i.d. is rejected, it is still not clear which of the two is the reason. For this reason we additionally apply Ramsey's RESET Test [335] and the McLeod and Li Test [286]. The Ramsey RESET-Test checks the null hypothesis of a correctly specified linear model by adding a certain number n of higher order fitted terms. If the coefficients of these terms are significantly different from zero, it can be inferred that the linear model is not good enough due to existing additive nonlinear dynamics. In contrast, the McLeod and Li Test whose results have been separately discussed above, is useful for detecting multiplicative nonlinearities. For EUR/GBP and EUR/JPY, the test results on squared ARMA residuals, as depicted in Table 8.5, seem to correspond. The McLeod Li test finds significant multiplicative nonlinearities at the 1% level. In addition, the Ramsey RESET test identifies additive nonlinearities at the 1% level, which is a particularly interesting result if we consider that such nonlinearities are generally thought to be of second order nature. Since these two results are consistent with the strong rejections of the null hypothesis at the 1% level by the BDS test statistic, it can be concluded that linear model residuals of the EUR/GBP and EUR/JPY series are significantly determined by first and second order nonlinearities.

After having removed multiplicative nonlinearities by the component GARCH models from (8.22) and (8.25), the BDS test statistic still rejects the null hypothesis of i.i.d. for both currency pairs. There are two possible explanations. First, although the McLeod Li Test results suggest the opposite, one might still think that

Table 8.5 Testing against Linearity

Null	Test	Input	Output		
			EUR/GBP	EUR/JPY	EUR/USD
Linearity	RESET	ARMA-, Residuals of Δy_t	$n = 1$: 2.81*	$n = 1$: 0.00	$n = 1$: 0.0015
			$n = 2$: 6.03***	$n = 2$: 6.64***	$n = 2$: 0.1662
			$n = 3$: 4.83***	$n = 3$: 4.68***	$n = 3$: 3.2273**
			$n = 4$: 4.04***	$n = 4$: 4.59***	$n = 4$: 2.4309**
i.i.d.	BDS	ARMA- Residuals of Δy_t	$m = 2$: 0.0099***	$m = 2$: 0.0076***	$m = 2$: -0.0007
			$m = 3$: 0.0172***	$m = 3$: 0.0178***	$m = 3$: 0.0001
			$m = 4$: 0.0235***	$m = 4$: 0.0248***	$m = 4$: 0.0014
			$m = 5$: 0.0268***	$m = 5$: 0.0288***	$m = 5$: 0.0025
		GARCH(1,1)- Residuals of Δy_t	$m = 2$: 0.0101***	$m = 2$: 0.0076***	$m = 2$: -0.0007
			$m = 3$: 0.0175***	$m = 3$: 0.0178***	$m = 3$: 0.0002
			$m = 4$: 0.0237***	$m = 4$: 0.0248***	$m = 4$: 0.0014
			$m = 5$: 0.0271***	$m = 5$: 0.0288***	$m = 5$: 0.0025

The columns from left to right denote the null hypothesis, name of the test applied, time series input, and the test statistic output per currency pair. *, **, and *** indicate rejection of the null hypothesis at the 10%, 5%, and 1% significance level. n denotes the number of fitted terms included in the test regression. m denotes the number of correlation dimensions for which the test statistic is computed. For Ramsey's RESET, the test statistics from the F-test are provided

the component GARCH models are prone to misspecification. Consequently, the rejection of the null would be due to excess multiplicative nonlinearities. Hsieh [197] examined the daily foreign exchange rates versus the USD for five major currencies – the GBP, the CAD, the DEM, the JPY, and the CHF. All five of these time series exhibit highly significant BDS statistics even after autocorrelation effects and linear holiday and day-of-the-week effects have been filtered out, thereby indicating the existence of strong nonlinear dependencies within these data series. Hsieh further finds that a GARCH(1,1) model with either a Student's t or a generalized error distribution can describe the Canadian Dollar and the Swiss Franc exchange rates very well and the Deutsche Mark exchange rate reasonably well. While a GARCH(1,1) model can also account for most of the nonlinear serial dependencies within the GBP and the JPY exchange rates, such GARCH models do not fit either of these time series very well. Similarly, [198] finds for value-weighted, size-decile portfolios of weekly stock returns from 1963 to 1987 that these returns also exhibit nonlinear serial dependencies and that conditional heteroskedasticity could be the source of these nonlinearities, but that none of the ARCH models seem to adequately describe these data. However, we argue that based on our previous analyses, the presence of additive nonlinearities is more likely. Thus, although it is possible for exchange rates to be linearly uncorrelated and nonlinearly dependent ([197], p. 340), as is the case for the EUR/JPY time series, it is not only the changing of variances that is responsible for the rejection of i.i.d. in exchange rate changes ([197], p. 359), but also the changing of the mean. For EUR/USD, contradictive results were obtained. On the one hand, additive and multiplicative nonlinearities could be detected separately by the RESET test and the McLeod Li test. On the other hand, these results could not be confirmed by the BDS test which would imply that ARMA residuals are i.i.d. The reason for this phenomenon could be found in the power of the BDS test itself. According to [61], applying the BDS test to residuals from some preliminary estimation of a fitted model can be problematic since the resultant "nuisance parameter" problem affects the behavior of the BDS statistic in finite samples and leads to the BDS test's having an actual size greater than its nominal size. This problem is exacerbated and exists even for large samples when the residuals come from a GARCH-type model, in which case the BDS test will lack power to reject a false model.

8.2.6 Results

The purpose of above analysis was to examine the degree of randomness of EUR/GBP, EUR/JPY, and EUR/USD time series statistically, in order to justify our SVM approach to exchange rate forecasting. In supervised learning, it is assumed that a functional relationship is implicitly reflected within the input/output pairings. However, this assumption may be questionable if the output data does not contain structure, but instead is severely corrupted by noise. Then it may be difficult to identify a reliable functional relationship. It was shown that all three time series contain statistically significant structure. EUR/GBP exhibits both linear dependencies, as well as nonlinear dependencies in the first and second moment. EUR/JPY does not exhibit linear dependencies but is significantly determined by first and second order nonlinearities. For EUR/USD, linear dependencies could be detected, but contradictive test results were obtained when examining potential nonlinearities. Still it remains to be seen whether SVMs are able to exploit nonlinear relationships out-of-sample and how they compare to linear benchmark models.

Chapter 9
Support Vector Classification

9.1 Binary Classification Problem (BCP)

Computational geometry is the branch of computer science that studies algorithms for solving geometric problems. The input to a computational-geometry problem is typically a description of a set of geometric objects, such as a set of points, a set of line segments, or the vertices of a polygon. The output is a response to a query about these objects (such as whether any of the lines intersect), or even a new geometric object (such as a convex hull or a separating hyperplane). The problem under consideration is the linear separability problem [112]. In the special case of finding whether two sets of points in general space can be separated, the linear separability problem becomes the binary classification problem (BCP). The most general form of the BCP is the case of whether two sets of points in general space can be separated by k hyperplanes. This problem is referred to as the k-polyhedral separability problem which has been formulated by [288] as follows:

Problem 9.1. Given a sample of l points $X = (x_1, \ldots, x_l)'$, a partition of this sample into two disjoint sets $S_1 = \{\pi_1, \ldots, \pi_p\} \subset R^d$ and sets $S_2 = \{\rho_1, \ldots, \rho_q\} \subset R^d$, and an integer k, recognize whether there exist k hyperplanes (that is k nonzero vectors $w \in R^d$ and k numbers b) $H_j = \{z : z^T w_j = b_j\}(w_j \in R^d, b_j \in R, j = 1, \ldots, k)$ that separate the sets S_1 and S_2 through a boolean formula as follows. Associate with each hyperplane H_j a boolean variable ξ_j. The variable ξ_j is true at a point z if $z^T w_j > b_j$ and false if $z^T w_j < b_j$. It is not defined at points lying on the hyperplane itself. A Boolean formula $\phi = \phi(\xi_1, \ldots, \xi_k)$ separates the sets S_1 and S_2 if ϕ is true at each of the points $\pi_1, \ldots, \pi_p$ and false at each of the points $\rho_1, \ldots, \rho_q$.

9.2 On the Computational Complexity of the BCP

The BCP is a well examined in computational geometry. Let us first consider the case of separating S_1 and S_2 with one hyperplane, i.e., $k = 1$.

C. Ullrich, *Forecasting and Hedging in the Foreign Exchange Markets,* Lecture Notes in Economics and Mathematical Systems 623, DOI: 10.1007/978-3-642-00495-7_9, © Springer-Verlag Berlin Heidelberg 2009

Problem 9.2. Given a sample of l points $X = (x_1, \ldots, x_l)'$, and a partition of this sample into two disjoint sets $S_1 = \{\pi_1, \ldots, \pi_p\} \subset R^d$ and sets $S_2 = \{\rho_1, \ldots, \rho_q\} \subset R^d$, recognize whether there exist a hyperplane (that is a nonzero vector $w \in R^d$ and one number b) $H = \{z \in R^d : w^T z = b\}$ (characterized by a nonzero vector $w \in R^d$, and a scalar b) that separates the sets S_1 and S_2 in the sense that for each point $\pi_i \in S_1$, $(\pi_i)^T w_i < b$ and for each point $\rho_i \in S_2$, $(\rho_i)^T w_i > b$.

If $k = 1$, the BCP can be formulated as a linear programing problem and is therefore solvable in polynomial time. However, when two sets cannot be separated by one hyperplane, a natural problem is to find the minimum number of hyperplanes that is required for their separation. For instance, if $k = 1$ the problem can be stated as follows.

Problem 9.3. Given a sample of l points $X = (x_1, \ldots, x_l)'$, and a partition of this sample into two disjoint sets $S_1 = \{\pi_1, \ldots, \pi_p\} \subset R^d$ and sets $S_2 = \rho_1, \ldots, \rho_q\} \subset R^d$, recognize whether there exist two hyperplanes $H_1 = \{z : w_1^T z = b_1\}$ and $H_2 = \{z : w_2^T z = b_2\}(w_1, w_2 \in R^d; b_1, b_2 \in R)$ that separates the sets S_1 and S_2 in the sense expressed by the following conditions: (i) For each point $\pi_i \in S_1$, both $(\pi_i)^T w_1 < b_1$ and $(\pi_i)^T w_2 < b_2$, (ii) For each point $\rho_i \in S_2$, either $(\rho_i)^T w_1 > b_1$ or $(\rho_i)^T w_2 > b_2$.

If $k = 2$, the problem becomes *NP*-complete and therefore cannot be solved by a polynomial time algorithm. This can be proven by reducing the reversible satisfiability problem with six literals per clause (Reversible 6-SAT), which is known to be *NP*-complete, to the 2-BCP (see [288]). As a natural extension, the k-linear separability problem which generalizes the 2-linear separability problem must also be *NP*-complete.

Moreover, it has been proven that for every fixed k and d the k-linear separability problem in R^d can be solved in polynomial time. If two sets $S_1, S_2 \in R^d$ are separable with k hyperplanes, then there exist k pairs of complementary subsets $A_i, B_i \subset S_1 \cup S_2$ (that is $A_i \cup B_i = S_1 \cup S_2, i = 1, \ldots, k$) such that H_i separates A_i from B_i. It thus follows that the separating hyperplanes can be chosen from a finite set. Each of the candidate hyperplanes is determined by some set of at most $d + 1$ points, together with a choice of at most d equalities $w_j = \pm 1$. The number of such sets of at most $d + 1$ points is polynomial in the cardinality of $S_1 \cup S_2$. Thus, the number of combinations of k such sets is also polynomial. It follows that all the relevant configurations of hyperplanes can be enumerated in polynomial time. Furthermore, it takes polynomial time to check whether a given configuration actually separates S_1 from S_2 which establishes the proof.

Another important result on the computational complexity of the BCP has been given by [187] who proved that the following problem is *NP*-hard: given a set of labeled examples, find the hyperplane that minimizes the number of misclassified examples both above and below the hyperplane. This result implies that any method for finding the optimal split is likely to have exponential cost if we assume that $P \neq NP$. We find this proof very relevant in the present context since classifiers are usually judged by how many points they classify correctly. Unfortunately, it is computationally infeasible to enumerate all $2^d(l, d)^T$ distinct hyperplanes and choose the best. In order to cope with computational complexity, heuristic search methods have been employed [188, 303]. However, any nonexhaustive deterministic algorithm is prone to getting stuck in a local minimum. A second possibility is to define error measures for which the problem of finding optimal hyperplanes can be solved in polynomial time. For example, if one minimizes the sum of distances of misclassified examples, then the optimal solution can be found using linear programing methods.

9.3 Supervised Learning

We approach the task of solving the BCP via machine learning. The field of machine learning is concerned with the question of how to construct computer programs that automatically improve with experience [297]. More specifically, we concentrate on supervised learning which is also known as learning from examples.

Definition 9.1. Let $y = f(x)$ denote an unknown functional relationship of an input x and an output y. The goal is to learn the function f from a limited number of training examples. The examples of input/output functionality as expressed by a set

$$D = ((x_1, y_1), ..., (x_l, y_l)), \ x \in R^n, \ y \in \{-1, +1\} \tag{9.1}$$

are referred to as training data.

In supervised learning, it is therefore assumed that a functional relationship f also known as the target function is implicitly reflected within the input/output pairings. The estimate of the target function which is learnt or output by the learning algorithm is known as the solution of the learning problem. In the case of classification this function is referred to as the decision function h. The decision function partitions the underlying vector space into two sets, one for each class. The classifier will classify all the points on one side of the decision boundary as belonging to one class and all those on the other side as belonging to the other class. If the decision function is a hyperplane, then the classification problem is called linear, and the classes are called linearly separable. The decision function is chosen from a set of candidate functions, also known as hypotheses H, which map from the input space to the output domain. The choice of the set of hypotheses determines the hypothesis space and represents the first important concept of the learning strategy. The algorithm which takes the training data as input and selects a hypothesis from the hypothesis space is termed learning algorithm and represents the second important ingredient.

In order to obtain a reliable estimate of h by the learning algorithm, several aspects must be considered. First, the learnt function should be able to explain the examples, i.e., $h(x_1) = y_1, h(x_2) = y_2, h(x_l) = y_l$, as good as possible. Figure 9.1 shows that there is also a second reason. The depicted training examples are perfectly explained by the dotted function as opposed to the linear function. Nevertheless, one is inclined to have more confidence in the straight line since it is more likely to generalize better, i.e., work better on unseen examples.

Fig. 9.1 Linear and nonlinear
classifier

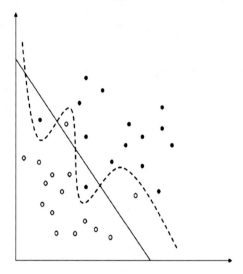

Fig. 9.2 Generalization error
as the sum of estimation and
approximation error

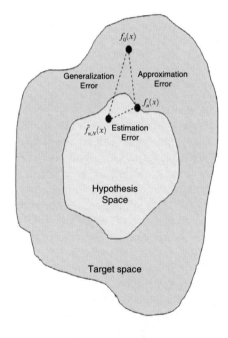

The complexity of the hypothesis space, i.e., the class of functions which is au-
thorized for solving the BCP, thus influences our confidence in the solution found.
The formalization of this insight provides the core of statistical learning theory
where the generalization ability is commonly quantified with respect to some error measure to the underlying function in the target space. This error measure is
termed generalization error and represents the sum of two errors which may arise
out of the following two cases as is visualized in Fig. 9.2:

- Case 1: The hypothesis space is smaller than the target space. As a consequence, the underlying function may lie outside the hypothesis space. A poor choice of the model space will result in a model mismatch which is measured by the approximation error.
- Case 2: The technique for selecting a model from the hypothesis space is not optimal. As a consequence, there is an error that is attributed to the learning procedure. This type of error is referred to as estimation error.

If we account for the importance of choosing an appropriate hypothesis space, the BCP can be summarized as follows. Given a set of decision functions $H \in \{\pm 1\}^X$ and a set of training examples according to (9.1) which are randomly generated according to a fixed unknown probability distribution $P(x,y)$, the goal is to choose a function h^* which best reflects the relationship between x and y for a set of test examples, drawn from the same probability distribution. The relationship between x and y is modeled by a probability distribution $P(x,y)$ which contains as a special case the possibility of a deterministic relationship $y = f(x)$. The best function is the one which reproduces the relationship between x and y best on average. This function can be found by minimizing the expected risk defined as the average probability of misclassified test examples

$$R[f] = \frac{1}{2} \int |f(x) - y| dP(x,y) \tag{9.2}$$

However, since the true probability distribution which generates the relationship between x and y is unknown, an inference procedure is required in order to at least approximate the function based on the observed training data.

9.4 Structural Risk Minimization

An intuitive and widespread inference procedure is empirical risk minimization (ERM), which considers the average probability of misclassified training examples in order to approximate (9.2)

$$R_{emp}[f] = \frac{1}{l} \sum_{i=1}^{l} |f(x_i) - y_i| \tag{9.3}$$

However, it has been argued that ERM is an incomplete inductive principle [295] since it is does not guarantee high generalization ability: an out-of-sample example generated by the same probability distribution does not necessarily lie on the dotted line from Fig. 9.1. In other words, it is problematic to infer from low empirical risk to low expected risk. In order to do so the capacity/complexity of the hypothesis space has to be examined which can be measured by the Vapnik–Chervonenkis- (VC-) Dimension.

The VC-dimension is an important property of a hypothesis space H and can be defined for a variety of hypotheses spaces. If a given set of l training points can be labeled in all possible 2^l ways, and for each labeling, a member of the hypothesis space can be found which correctly assigns those labels, the set of points is said to be shattered by that set of functions S_p.

Definition 9.2. The VC-dimension for the set of functions S_p is defined as the maximum number of training points that can be shattered by S_p.

If the VC-dimension is p, then there exists at least one set of p points that can be shattered, but in general it will not be true that every set of p points can be shattered. For instance, suppose that R^2 is the space in which the training data live and that for given separating lines, all points on the one side are assigned the class 1, and all points on the other side the class -1. According to Fig. 9.3, which is taken from [361], p. 10, there exist $2^3 = 8$ possibilities to divide 3 training points into 2 classes. While it is possible to find three points that can be shattered by this set of functions, it is not possible to find four. Thus the VC-dimension of the set of oriented lines in R^2 is $p = 3$. Hence p is referred to as the VC-dimension of a hypothesis space S_p, if and only if there exists a set of points $\{x_i\}_{i=1}^p$ such that these points can be separated in all 2^p possible configurations and that no set $\{x_i\}_{i=1}^q$ exists where $q > p$ satisfies this property.

The trade-off between ERM and hypothesis space capacity as measured by the VC-dimension is described by probabilistic bounds whose study is a major subject of statistical learning theory. The meaning of these bounds can be explained as follows: if it is possible to explain the training data (i.e., to keep the empirical risk low) by a simple model (i.e., a hypothesis space whose VC-dimension is small compared to the number of training examples), there is good reason for assuming that the true functional relationship has been found. Otherwise, if the training data can only be explained by a hypothesis space of higher VC-dimension, this is not the case, since the machine may have used its capacity to memorize the single training examples (overfitting) instead of learning a more compact underlying regularity. Accordingly, it cannot be expected that out-of-sample examples can be reliably classified. The principle of structural risk minimization (SRM) uses probabilistic bounds in order to minimize the expected risk by controlling both empirical risk and VC-dimension in order to find a function that generalizes well on new unseen examples [85, 404, 405]

$$\min_{S_h} R_{emp}[f] + \sqrt{\frac{p \ln\left(\frac{2l}{p} + 1\right) - \ln\left(\frac{\delta}{4}\right)}{l}} \qquad (9.4)$$

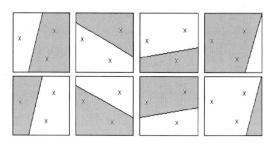

Fig. 9.3 Vapnik–Chervonenkis (VC-) dimension

In this sense, SRM adapts the complexity of the learning machine to the problem under consideration and is therefore superior to the traditional ERM principle [176] which has been used in many traditional neural network approaches.

In some cases, the VC-dimension is equal to the number of parameters of the hypothesis space which in turn is related to the dimensionality of the observations x_i. Therefore, the number of free parameters is sometimes considered as a measure of complexity or capacity of a hypothesis space. However, this is not always true with the negative consequence that the estimation of the capacity of a hypothesis space cannot be reduced to counting of parameters, but in fact may be a difficult mathematical problem. On the other hand, the advantage is that one may still hope to be capable of generalizing even if the data is high-dimensional. A class of learning algorithms which have confirmed these hopes in various areas are support vector machines (SVM), or more generally, kernel algorithms.

9.5 Support Vector Machines

The difficulty in estimating the capacity of hypothesis spaces is that measures such as the VC-Dimension are of combinatorial nature. This leads to a dilemma: good estimates exist particularly for simple (i.e., linear) hypothesis spaces. However, in order to analyze complex phenomena in the scientific domain, the goal is often to construct learning algorithms for nonlinear hypothesis spaces. SVMs are kernel algorithms which solve this dilemma in an elegant way by focusing on two major principles: in a first step, SVMs map training examples in a high-dimensional feature space in order to construct a hyperplane which separates the two classes in a second step. This will be explained in more detail in the following two subsections.

9.5.1 Learning in Feature Space

Definition 9.3. SVMs are kernel algorithms which apply a mapping ϕ from the original input space X into a high-dimensional feature space F

$$X = ((x_1, y_1), \ldots, (x_l, y_l)) \rightarrow F = \phi(X) = (\phi(x_1, y_1), \ldots, \phi(x_l, y_l)) \qquad (9.5)$$

where F is a Hilbert space, i.e., a complete vector space provided with an inner product.

Figure 9.4 shows a conceptual example of such a feature mapping from a two-dimensional input space to a two dimensional feature space ([85], p. 28). The data cannot be separated by a linear function in the input space. However, the transformed feature vectors $\phi(x_i, y_i)$ can be separated by a linear function in feature space. The decision boundary which is linear in F corresponds to a nonlinear decision boundary in X.

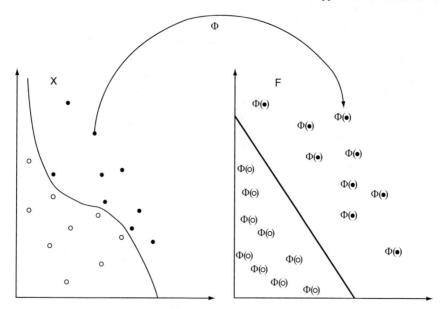

Fig. 9.4 Mapping from input space to feature space

A feature map can therefore simplify the classification task. But how can one identify the possibly best feature map for a given classification task? In the machine learning community, the task of choosing the most suitable representation of the data is known as feature selection. The objective of feature selection is to identify the smallest set of features that still conveys the essential information contained in the original attributes. This requires reducing the dimensionality of a given dataset which is beneficial since both the computational and generalization performance can degrade as the number of features grow. However, transforming the vectors in the training set D into a higher-dimensional space may incur computational problems. For very large training sets, the high dimensionality of F makes it very expensive, both in terms of memory and time, to represent the feature vectors $\phi(x_i, y_i)$ corresponding to the training vectors (x_i, y_i).

9.5.2 Kernel Functions

For a formal definition of a kernel function we refer to [85], p. 30.

Definition 9.4. A kernel is a symmetric function $K : X \times X \to R$ so that for all x_i and x_j in X, $K\langle \phi(x_i), \phi(x_j) \rangle$ where ϕ is a (nonlinear) mapping from the input space X into the Hilbert space F provided with the inner product $K\langle ., . \rangle$.

The inner product, therefore, does not need to be evaluated in the feature space which provides a way of addressing the curse of dimensionality and has been referred to as the so called *kernel trick* [360]. However, the computation is still critically dependent upon the number of training patterns.

A characterization of when a function $K\langle x,z \rangle$ is a kernel has been provided by Mercer's theorem [289] which is founded on reproducing Kernel Hilbert spaces (RKHS) (see for instance [11, 412]):

Theorem 9.1. (Mercer's Theorem). *Let X be a compact subset of R^n. Suppose K is a continuous symmetric function such that the integral operator $T_K : L_2(X) \to L_2(X)$*

$$(T_K f)(.) = \int_X K(.,X)f(x)dx \qquad (9.6)$$

is positive, that is

$$\int \int K(x,z)f(x)f(z)dxdz \geq 0 \qquad (9.7)$$

for all $f \in L_2(X)$. Then we can expand $K\langle x,z \rangle$ in a uniformly convergent series (on $X \times X$) in terms of T_K's eigenfunctions $\phi_j \in L_2(X)$, normalized in such a way that $\|\phi_j\|_{L2}$, and positive associated eigenvalues $\lambda_j \geq 0$,

$$K(x,z) = \sum_{j=1}^{\infty} \lambda_j \phi_j(x)\phi_j(z) \qquad (9.8)$$

Hence, an inner product in feature space has an equivalent kernel in input space, $K(x,z) = \langle \phi(x), \phi(z) \rangle$ if K is a symmetric positive definite function that satisfies Mercer's conditions as represented by (9.6) and (9.7).

The main attraction of kernel functions is not only the fact that the application of kernel algorithms has proven to be very successful and includes world records on famous benchmark data [92, 415]. Kernels are also very interesting from a theoretical perspective since they deal with three fundamental problems of empirical inference:

- Data representation: a kernel $K(x,z)$ induces an imbedding of the data in the vector space
- Similarity: a kernel $K(x,z)$ can be perceived as a (nonlinear) measure of similarity which can be used for comparing data points
- A-priori-knowledge: the solutions of kernel learning algorithms may be generally expressed as kernel developments in the training data. For this reason, a kernel $K(x,z)$ parameterizes the hypothesis space which contains the solution, i.e., the knowledge which enters the learning process jointly with the training data.

9.5.3 Optimal Separating Hyperplane

Once the training examples are mapped into a high-dimensional feature space, the goal is to construct a hyperplane in order to separate the two classes. There exist many possible linear classifiers that can separate the data and therefore solve

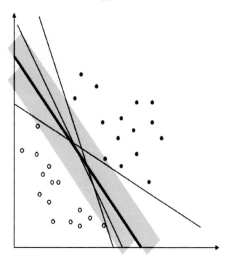

Fig. 9.5 Optimal separating
hyperplane

the BCP. However, there is only one that maximizes the distance between itself
and the nearest data point of each class. This linear classifier is termed the optimal
separating hyperplane or maximum margin classifier. By explicitly maximizing the
margin, the advantage of this sort of classifier is that it reduces the function capac-
ity/complexity and therefore, minimizes bounds on the generalization error. It has
maximal stability because it ensures the lowest probability of misclassification. This
can be explained by using geometric arguments paired with the ideas from Sect. 9.4.
Consider the example in Fig. 9.5 which illustrates the binary classification toy problem of separating white from black dots.

The data can be separated in different ways with margins indicated by the slim
lines. The optimal separating hyperplane is represented by the fat line. Figure 9.5
also shows how a linear discriminant that separates two classes with a small margin
has more capacity to fit the data than one with a large margin. The capacity of a
linear discriminant is a function of the margin of separation. A hyperplane with a
small margin has more capacity since it can take many possible orientations and still
strictly separate all the data. On the other hand, a hyperplane with a large margin
has limited flexibility to separate the data and has, therefore, a lower capacity than
a small margin one. In general, it is thought that the capacity of a linear function is
being determined by the number of variables. However, the size of the margin is not
directly dependent on the dimensionality of the data. Thus, if the margin is fat, then
the capacity of a function as measured by its VC-dimension can be low even if the
number of variables is very high. As a consequence, problems caused by overfitting
of high-dimensional data are greatly reduced and good performance, even for very
high-dimensional data, can be expected.

Formally, the set of training examples D belonging to two different classes is said
to be optimally separated by the hyperplane

$$\langle w, x \rangle + b = 0 \qquad\qquad (9.9)$$

Fig. 9.6 Separating hyperplane (w, b) for a two dimensional training set

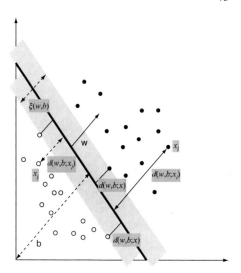

if it is separated without error and the distance between the closest vector to the hyperplane is maximal. The parameters (w, b) determine the weight vector and bias, and must be learned from the data by a given learning methodology. A geometric interpretation is given in Fig. 9.6 where the hyperplane is the dark line with the positive region (black dots) above and the negative region (white dots) below. The vector w defines a direction perpendicular to the hyperplane, while varying the value of b moves the hyperplane parallel to itself.

The functional margin of a training example (x_i, y_i) with respect to a hyperplane (w, b) is described by

$$d_F(w, b; x_i) = y_i(\langle w, x_i \rangle + b) \tag{9.10}$$

The distribution of the margins of the examples in a training set S is the functional margin distribution of (w, b). The minimum of the functional margin distribution is called the functional margin of (w, b) with respect to S and is given by

$$d_F = \min_i d_F(w, b; x_i) \tag{9.11}$$

Equation 9.9 can be simplified without loss of generality by considering a canonical hyperplane [404], where the parameters w and b are constrained by

$$\min_i |\langle w, x_i \rangle + b| = 1 \tag{9.12}$$

It states that the norm of the weight vector should be equal to the inverse of the distance of the nearest point in the data set to the hyperplane. A separating hyperplane in canonical form must satisfy the following constraints

$$d(w, b; x_i) = y_i(\langle w, x_i \rangle + b) \geq 1 \tag{9.13}$$

According to above, the distance $d(w,b;x_i)$ of a point x_i from the canonical hyperplane (w,b) is termed geometric margin of x_i from (w,b). The distribution of the geometric margins of the examples in a training set S is the geometric margin distribution of (w,b). The minimum of the functional margin distribution is called the geometric margin of (w,b) with respect to S and is given by

$$d(w,b;x) = \min_i \frac{1}{\|w\|} d(w,b;x_i) \tag{9.14}$$

and measures the Euclidean distances of the points from the decision boundary in the input space. Thus, the geometric margin d will equal the functional margin if the weight vector is a unit vector. Figure 9.6 illustrates the geometric margin at two points x_i and x_j with respect to a hyperplane in two dimensions. The maximum geometric margin over all hyperplanes is referred to as the margin ξ of a training set S. A hyperplane realizing this maximum is known as a maximum margin hyperplane. The optimal hyperplane is given by maximizing the margin

$$\begin{aligned}
\xi(w,b) &= \min_{x_i:y_i=-1} d(w,b;x_i) + \min_{x_i:y_i=-1} d(w,b;x_i) \\
&= \min_{x_i:y_i=-1} \frac{(|\langle w,x_i \rangle + b|)}{\|w\|} + \min_{x_i:y_i=+1} \frac{(|\langle w,x_i \rangle + b|)}{\|w\|} \\
&= \frac{1}{\|w\|} \left(\min_{x_i:y_i=-1} |\langle w,x_i \rangle + b| + \min_{x_i:y_i=+1} |\langle w,x_i \rangle + b| \right) \\
&= \frac{2}{\|w\|}
\end{aligned} \tag{9.15}$$

Hence, the hyperplane that optimally separates the data is the one that minimizes

$$\Phi(w) = \frac{1}{2} \|w\|^2 \tag{9.16}$$

It is independent of b because provided (9.13) is satisfied (i.e., it is a separating hyperplane), changing b will move it in the normal direction to itself. Accordingly, the margin remains unchanged but the hyperplane is no longer optimal in that it will be nearer to one class than the other. To consider how minimizing (9.16) is equivalent to implementing the SRM principle, suppose that the following bound holds,

$$\|w\| < A \tag{9.17}$$

Then from (9.13) and (9.14),

$$d(w,b;x) \geq \frac{1}{A} \tag{9.18}$$

Accordingly, the hyperplanes cannot be nearer than $(1/A)$ to any of the data points. This reduces the amount of possible hyperplanes and hence, the capacity. The VC-dimension p, of the set of canonical hyperplanes in n dimensional space is bounded by

$$p \leq min\left[R^2 A^2, n\right] + 1 \tag{9.19}$$

where R is the radius of a hypersphere enclosing all the data points. Hence, minimizing (9.16) is equivalent to minimizing an upper bound on the VC-dimension. The solution to the optimization problem of (9.16) under the constraints of (9.13) is given by the saddle point of the Lagrange function [296].

$$\Phi(w,b,\alpha) = \frac{1}{2}\|w\|^2 - \sum_{i=1} \alpha_i \left(y_i[\langle w, x_i \rangle + b] - 1 \right) \qquad (9.20)$$

where α are the Lagrange multipliers. (9.20) has to be minimized with respect to w and b, and maximized with respect to $\alpha \geq 0$. Classical Lagrangian duality enables the primal problem, (9.20), to be transformed to its dual problem, which is easier to solve. The dual problem is given by

$$\max_\alpha W(\alpha) = \max_\alpha \left(\min_{w,b} \Phi(w,b,\alpha) \right) \qquad (9.21)$$

The minimum with respect to w and b of the Lagrange function Φ is given by

$$\frac{\partial \Phi}{\partial b} = 0 \Rightarrow \sum_{i=1}^l \alpha_i y_i = 0 \qquad (9.22)$$

$$\frac{\partial \Phi}{\partial w} = 0 \Rightarrow w = \sum_{i=1}^l \alpha_i y_i x_i$$

Hence, from (9.20)–(9.22), the dual problem is

$$\max_\alpha W(\alpha) = \max_\alpha -\frac{1}{2} \sum_{i=1}^l \sum_{j=1}^l \alpha_i \alpha_j y_i y_j \langle x_i, x_j \rangle + \sum_{k=1}^l \alpha_k \qquad (9.23)$$

The solution to the problem is given by

$$\alpha^* = \arg\min_\alpha -\frac{1}{2} \sum_{i=1}^l \sum_{j=1}^l \alpha_i \alpha_j y_i y_j \langle x_i, x_j \rangle - \sum_{k=1}^l \alpha_k$$

w.r.t. $\qquad\qquad (9.24)$

$$\alpha_i \geq 0, \ i = 1,\ldots,l$$

$$\sum_{j=1}^l \alpha_j y_j = 0$$

where the sign of the coefficient of x_i is given by the classification of y_i, the α_i are positive values proportional to the number of times misclassification of x_i has caused the weight to be updated. Points that have caused fewer mistakes will have smaller α_i, whereas difficult points will have larger values. Thus, in the case of nonseparable data, the coefficients of misclassified points must grow indefinitely.

Solving (9.24) determines the Lagrange multipliers. The optimal separating hyperplane is then given by

$$w^* = \sum_{i=1}^{l} \alpha_i y_i x_i \tag{9.25}$$

$$b^* = -\frac{1}{2}\langle w^*, x_r + x_s \rangle$$

where x_r and x_s are any support vector from each class satisfying

$$\alpha_r, \alpha_s > 0, \ y_r = -1, \ y_s = +1 \tag{9.26}$$

The hard classifier is then

$$f(x) = \text{sgn}(\langle w^*, x \rangle + b) \tag{9.27}$$

Alternatively, a soft classifier may be used which linearly interpolates the margin

$$f(x) = h(\langle w^*, x \rangle + b), \text{ where } h(z) = \begin{cases} -1 & : & z < -1 \\ z & : & -1 \leq z \leq 1 \\ 1 & : & z > 1 \end{cases} \tag{9.28}$$

This may be more appropriate than the hard classifier of (9.27) in cases where the training data is linearly separable except for some exceptions. The soft classifier produces a real valued output between -1 and 1 when the classifier is queried within the margin, where no training data resides. From the Karush–Kuhn–Tucker (KKT) conditions

$$\alpha_i(y_i[\langle w, x_i \rangle + b] - 1) = 0, \ i = 1, \dots, l \tag{9.29}$$

and hence only the points x_i which satisfy

$$y_i[\langle w, x_i \rangle + b] = 1 \tag{9.30}$$

will have nonzero Lagrange multipliers. These points are termed support vectors (SV). If the data is linearly separable all the SV will lie on the margin and hence the number of SV can be very small. Consequently, the hyperplane is determined by a small subset of the training set; the other points could be removed from the training set and recalculating the hyperplane would produce the same answer. Hence, SVM can be used to summarize the information contained in a data set by the SV produced. If the data is linearly separable the following equality will hold

$$\|w\|^2 = \sum_{i=1}^{l} \alpha_i = \sum_{i \in SV} \alpha_i = \sum_{i \in SV} \sum_{j \in SV} \alpha_i \alpha_j y_i y_j \langle x_i, x_j \rangle \tag{9.31}$$

Hence, from (9.19) the VC dimension of the classifier is bounded by

$$p \leq \min \left[R^2 \sum_{i \in SV}, n \right] + 1 \tag{9.32}$$

and if the training data X is normalized to lie in the unit hypersphere, the VC-dimension of the classifier is bounded by

$$p \leq 1 + \min\left[\sum_{i \in SV}, n\right] \qquad (9.33)$$

9.5.4 Generalized Optimal Separating Hyperplane

So far the discussion has been restricted to the case where the training data is (strictly) linearly separable. However, this must not be the case. Figure 9.7 shows that there is no linear classifier that is able to separate the two classes without error in the input space. Linear separation of input points does, therefore, not work well: a reasonably sized margin requires misclassifying four points.

There are two approaches to generalizing the problem, which are dependent upon prior knowledge of the problem and an estimate of the noise on the data. In the case where it is expected or possibly even known that a hyperplane can correctly separate the data, it is appropriate to focus on a different measure of the margin distribution, an additional cost function associated with misclassification, which generalizes the notion of margin to account for a more global property of the training sample.

Alternatively, a more complex function can be used to describe the boundary (see Sect. 9.5.2). To enable the optimal separating hyperplane method to be generalized, [83] introduced a nonnegative margin slack variable ζ_i of an example (x_i, y_i) with respect to the hyperplane (w, b) as

$$\zeta_i = \zeta((x_i, y_i), (w, b), d(w, b; x))$$
$$= \max(0, d(w, b; x) - y_i(\langle w, x_i \rangle + b)) \qquad (9.34)$$

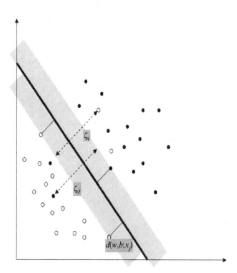

Fig. 9.7 Generalized optimal separating hyperplane

given a fixed target margin $d(w,b;x)$. Informally, this quantity measures how much a point fails to have a margin of $d(w,b;x)$ from the hyperplane. If $\zeta_i > d(w,b;x)$, then x_i is misclassified by (w,b). The norm $\|\zeta\|_2$ measures the amount by which the training set fails to have a margin $d(w,b;x)$ and takes into account any misclassifications of the training data. Figure 9.7 shows the size of the margin slack variables for two misclassified points for a hyperplane with unit norm. All of the other points in the figure have their slack variable equal to zero since they have a positive margin of more than $d(w,b;x)$.

The slack variables are associated with a penalty function

$$F_\sigma(\zeta) = \sum_i \zeta_i^\sigma, \ \sigma > 0 \tag{9.35}$$

where ζ_i are a measure of the classification errors. The optimization problem is to minimize the classification error as well as minimizing the bound on the VC-dimension of the classifier. The constraints of (9.13) are modified for the nonseparable case to

$$y_i(\langle w, x_i \rangle + b) \geq 1 - \zeta_i \tag{9.36}$$

where $\zeta_i \geq 0$. The generalized optimal separating hyperplane is determined by the vector w, that minimizes the function

$$\Phi(w, \zeta) = \frac{1}{2}\|w\|^2 + C\sum_i \zeta_i \tag{9.37}$$

where C is a given value subject to the constraints of (9.36). The solution to the optimization problem of (9.37) under the constraints of (9.36) is given by the saddle point of the Lagrange function [296]

$$\Phi(w,b,\alpha,\zeta,\beta) = \frac{1}{2}\|w\|^2 + C\sum_i \zeta_i$$

$$- \sum_{i=1}^{l} \alpha_i(y_i[w^T x_i + b] - 1 + \zeta_i) \tag{9.38}$$

$$- \sum_{j=1}^{l} \beta_i \zeta_i$$

where α, β are the Lagrange multipliers. The Lagrangian has to be minimized with respect to w, b, x, and maximized with respect to α, β. As before, classical Lagrangian duality enables the primal problem, (9.37), to be transformed to its dual problem. The dual problem is given by,

$$\max_\alpha W(\alpha,\beta) = \max_{\alpha,\beta} W\left(\min_{w,b,\zeta} \Phi(w,b,\alpha,\zeta,\beta)\right) \tag{9.39}$$

The minimum with respect to w, b, ζ of the Lagrangian F is given by

$$\frac{\partial \Phi}{\partial b} = 0 \Rightarrow \sum_{i=1}^{l} \alpha_i y_i = 0$$

$$\frac{\partial \Phi}{\partial w} = 0 \Rightarrow w = \sum_{i=1}^{l} \alpha_i y_i x_i \qquad (9.40)$$

$$\frac{\partial \Phi}{\partial \zeta} = 0 \Rightarrow \alpha_i + \beta_i = C$$

Hence, from (9.38)–(9.40), the dual problem becomes

$$\max_{\alpha} W(\alpha) = \max_{\alpha} \ -\frac{1}{2} \sum_{i=1}^{l} \sum_{j=1}^{l} \alpha_i \alpha_j y_i y_j \langle x_i, x_j \rangle + \sum_{k=1}^{l} \alpha_k \qquad (9.41)$$

and the solution to the problem is given by,

$$\alpha^* = \arg \ \min_{\alpha} \ \frac{1}{2} \sum_{i=1}^{l} \sum_{j=1}^{l} \alpha_i \alpha_j y_i y_j \langle x_i, x_j \rangle - \sum_{k=1}^{l} \alpha_k$$

w.r.t.   (9.42)

$$0 \le \alpha_i \le C, \ i = 1, \ldots, l$$

$$\sum_{j=1}^{l} \alpha_j y_j = 0$$

9.5.5 Generalization in High Dimensional Feature Space

There are some restrictions on the nonlinear mapping that can be employed (see Sect. 9.5.1). The optimization problem of (9.42) becomes,

$$\alpha^* = \arg \ \min_{\alpha} \ \frac{1}{2} \sum_{i=1}^{l} \sum_{j=1}^{l} \alpha_i \alpha_j y_i y_j K \langle x_i, x_j \rangle - \sum_{k=1}^{l} \alpha_k$$

w.r.t.   (9.43)

$$0 \le \alpha_i \le C, \ i = 1, \ldots, l$$

$$\sum_{j=1}^{l} \alpha_j y_j = 0$$

$$(9.44)$$

where $K(x, x')$ is the kernel function performing the nonlinear mapping into feature space, and the constraints are unchanged. Solving Problem 9.43 determines the Lagrange multipliers, and a hard classifier implementing the optimal separating hyperplane in the feature space is given by

$$f(x) = \text{sgn}\left(\sum_{i \in SV} \alpha_i y_i K(x_i, x) + b\right)$$
(9.45)

where

$$\langle w^*, x \rangle = \sum_{i=1}^{l} \alpha_i y_i K(x_i, x)$$
(9.46)

$$b^* = -\frac{1}{2} \sum_{i=1}^{l} \alpha_i y_i [K(x_i, x_r) + K(x_i, x_r)]$$

The bias is computed using two support vectors, but can be computed using all the SV on the margin for stability [406]. If the kernel contains a bias term, the bias can be accommodated within the kernel, and hence the classifier is

$$f(x) = \text{sgn}\left(\sum_{i \in SV} \alpha_i K(x_i, x)\right)$$
(9.47)

Many employed kernels have a bias term and any finite Kernel can be made to have one [158]. This simplifies the optimization problem by removing the equality constraint of (9.43). However, even though SVMs implement the SRM principle and hence can generalize well, a choice of the kernel function is necessary to produce a classification boundary that is topologically appropriate. It is always possible to map the input space into a dimension greater than the number of training points and produce a classifier with no classification errors on the training set. However, this will generalize badly since the function learnt is very complex.

Chapter 10
Description of Empirical Study and Results

In the following two sections we will focus on the task of predicting a rise (labeled "+1") or fall (labeled "−1") of daily EUR/GBP, EUR/JPY, and EUR/USD exchange rate returns. To predict that the level of the EUR/USD, for instance, is close to the level today, is trivial. On the contrary, to determine if the market will rise or fall is much more interesting for a currency trader whose primary focus is to buy the base currency if the exchange rate went down and to sell the base currency if the exchange rate went up.

All of the following experiments were performed on a PC equipped with an Intel Pentium M Processor 750 and running at 1,866 MHz with 1,024 MB of system memory under the Windows XP operating system. The simulations were programed in the R environment [201], an open source and high-level programing language that provides powerful tools for statistical analysis. The R packages *1071* [70] and *kernlab* [220] were adopted for the SVM model fitting. These packages use the SMO algorithm that is implemented by the LIBSVM tool [70].

10.1 Explanatory Dataset

Exchange rate forecasting methods generally fall into one of the following three categories: fundamental forecasts, technical forecasts, or mixed forecasts. The forecasting approach that we adopted is what we call a statistical or purely data driven approach that borrows from both fundamental and technical analysis principles. It is fundamental since it considers relationships between the exchange rate and other exogenous financial market variables. However, our approach also has a technical component: it is somewhat irrational in a financial context since it depends heavily on the concepts of statistical inference. The fact that the input data is publicly available lets us challenge the EMH in its semi-strong version. The procedure of obtaining an exploratory dataset was divided into two phases according to [327]:

C. Ullrich, *Forecasting and Hedging in the Foreign Exchange Markets,* Lecture Notes in Economics and Mathematical Systems 623, DOI: 10.1007/978-3-642-00495-7_10, © Springer-Verlag Berlin Heidelberg 2009

1. Specifying and collecting a large amount of data
2. Reducing the dimensionality of the dataset by selecting a subset of that data for efficient training (feature extraction)

Since there is a trade-off between accuracy as represented by the entire dataset and the computational overheads of retaining all parameters without application of feature extraction/selection techniques, the data selection procedure is also referred to as the *curse of dimensionality* which was first noted by [32]. The merit of feature extraction is to avoid multicollinearity, a problem that is common to all sorts of regression models. If multicollinearity exists, explanatory variables have a high degree of correlation between themselves meaning that only a few important sources of information in the data are common to many variables. In this case, it may not be possible to determine their individual effects.

10.1.1 Phase One: Input Data Selection

The obvious place to start selecting data, along with the EUR/GBP, EUR/JPY, and EUR/USD is with the other leading traded exchange rates. In addition, we selected related financial market data, including stock market price indices, 3-month interest rates, 10-year government bond yields and spreads, the price of Brent Crude oil, and the prices of silver, gold, and platinum. Due to the bullish commodity markets, we also decided to include daily prices of assorted metals being traded on the London Metal Exchange, as well as agricultural commodities. Fundamental variables hardly play a role in daily FX movements and were disregarded. The reason is that such macroeconomic data are collected on a monthly, quarterly, or annual basis only and that it is thus believed to be more suitable for longer-run exchange rate predictions. All data were obtained from Bloomberg.

All the series span a 7-year time period from 1 January 1997 to 31 December 2004, totaling 2,349 trading days. The data is divided into two periods: the first period runs from 1 January 1997 to 31 August 2003 (1,738 observations), is used for model estimation and is classified in-sample. The second period, from 1 September 2003 to 31 December 2004 (350 observations), is reserved for out-of-sample forecasting and evaluation. Missing observations on bank holidays were filled by linear interpolation, which is a quick and easy approach to smoothen out these irregularities.

10.1.2 Phase Two: Dimensionality Reduction

Having collected an extensive list of candidate variables, the explanatory viability of each variable was evaluated. The aim was to remove those input variables that do not contribute significantly to model performance. For this purpose, we took a two-step procedure.

First, pair-wise Granger Causality tests [170] with lagged values until $k = 20$ were performed on stationary $I(1)$ candidate variables. The Granger approach to the question of whether an independent variable x causes a dependent variable y is to see how much of the current y can be explained by past values of y and then to see whether adding lagged values of x can improve the explanation. The variable y is said to be Granger-caused by x if x helps in the prediction of y, or equivalently if the coefficients on the lagged x's are statistically significant. The pair-wise Granger causality test is carried out by running bivariate regressions of the form

$$y_t = \alpha_0 + \alpha_1 y_{t-1} + \ldots + \alpha_l y_{t-l} + \beta_1 x_{t-1} + \ldots + \beta_l x_{t-l} + \varepsilon_t \qquad (10.1)$$
$$x_t = \alpha_0 + \alpha_1 x_{t-1} + \ldots + \alpha_l x_{t-l} + \beta_1 y_{t-1} + \ldots + \beta_l y_{t-l} + u_t$$

for all possible pairs of (x, y) series in the group. In general it is better to use more rather than fewer lags in the test regressions, since the theory is couched in terms of the relevance of all past information. It is also important that the lag length picked corresponds to reasonable beliefs about the longest time over which one of the variables could help predict the other. The reported F-statistics are the Wald statistics for the joint hypothesis:

$$\beta_1 = \beta_2 = \ldots = \beta_l = 0 \qquad (10.2)$$

for each equation. The null hypothesis is that x does not Granger-cause y in the first regression and that y does not Granger-cause x in the second regression. The major advantage of the Granger causality principle is that it is able to distinguish causation from correlation. Hence the known problem of spurious correlations can be avoided [173]. Still, it is important to note that the statement "x Granger causes y" does not imply that y is the effect or the result of x. Granger causality measures precedence and information content but does not by itself indicate causality in the more common use of the term. Tables 10.1–10.3 summarize the variables that were found to be relevant for explaining the single exchange rates as can be seen by the values of the F-statistic which leads at least to a rejection of the null hypothesis at the 10% significance level. Note that only the first regression from (10.1) has to be considered in our context.

We find that EUR/GBP is Granger-caused by 11 variables, namely

- EUR/USD, JPY/USD, and EUR/CHF exchange rates
- IBEX, MIB30, CAC, and DJST stock market indices
- The prices of platinum (PLAT) and nickel (LMNIDY)
- Ten-year Australian (GACGB10) and Japanese (GJGB10) government bond yields

Further, we identify ten variables that significantly Granger-cause EUR/JPY, namely

- EUR/CHF exchange rate
- IBEX stock market index
- The price of silver (SILV)
- Australian 3-month interest rate (AU0003M)

Table 10.1 Granger causality test results for EUR/GBP ($k = 20$)

Null hypothesis	F-Statistic
EUR/USD does not Granger cause EUR/GBP	1.4366*
GACGB10 does not Granger cause EUR/GBP	1.4297*
GJGB10 does not Granger cause EUR/GBP	1.4250*
IBEX does not Granger cause EUR/GBP	1.7812**
JPY/USD does not Granger cause EUR/GBP	1.6432**
LMNIDY does not Granger cause EUR/GBP	1.4927*
MIB30 does not Granger cause EUR/GBP	1.9559***
PLAT does not Granger cause EUR/GBP	1.4732*
CAC does not Granger cause EUR/GBP	1.7350**
CHF/USD does not Granger cause EUR/GBP	1.6131**
DJST does not Granger cause EUR/GBP	1.8096**

Table 10.2 Granger causality test results for EUR/JPY ($k = 20$)

Null hypothesis	F-Statistic
GACGB10 does not Granger cause EUR/JPY	1.6036**
GDBR10 does not Granger cause EUR/JPY	1.5556*
GJGB10 does not Granger cause EUR/JPY	1.7328**
GSWISS10 does not Granger cause EUR/JPY	1.4303*
GT10 does not Granger cause EUR/JPY	1.4502*
IBEX does not Granger cause EUR/JPY	1.5328*
SILV does not Granger cause EUR/JPY	1.5305*
AU0003M does not Granger cause EUR/JPY	2.1236***
BPSW10-GUKG10 does not Granger cause EUR/JPY	1.5779**
EUR/CHF does not Granger cause EUR/JPY	2.1408***

Table 10.3 Granger causality test results for EUR/USD ($k = 20$)

Null hypothesis	F-Statistic
LMSNDY does not Granger cause EUR/USD	1.4905*
LMZSDY does not Granger cause EUR/USD	1.5742*
KC1 does not Granger cause EUR/USD	1.6184**
LMCADY does not Granger cause EUR/USD	1.5256*
SPX does not Granger cause EUR/USD	2.1828***
AUD/USD does not Granger cause EUR/USD	1.4324*
CO1 does not Granger cause EUR/USD	1.8744**

- Australian (GACGB10), German (GDBR10), Japanese (GJGB10), Swiss (GSWISS10), and US (GT10) government bond yields
- UK bond spreads (BPSW10-GUKG10)

For EUR/USD, Granger causality tests yield seven significant explanatory variables:

- AUD/USD exchange rate
- SPX stock market index

- The prices of copper (LMCADY), tin (KMSNDY), zinc (LMZSDY), coffee (KC1), and cocoa (CO1)

Second, we carried out linear principal component analysis (PCA) on Granger caused explanatory datasets in order to check for computational overheads. PCA is generally considered as a very efficient method for dealing with the problem of multicollinearity. It allows for reducing the dimensionality of the underlying dataset by excluding highly intercorrelated explanatory variables. This results in a meaningful input for the learning machine.

The procedure is based on an eigenvalue and eigenvector analysis of $V = X'X/T$, the $k \times k$ symmetric matrix of correlations between the variables in X. Each principal component is a linear combination of these columns where the weights are chosen in such a way that

- The first principal component explains the greatest amount of the total variation in X, the second component explains the greatest amount of the remaining variation, and so on
- The principal components are uncorrelated with each other

The results of the PCAs for the respective exchange rates are given by Tables 10.4–10.6. The column headed by "C1" and "V1" corresponds to the first principal component, "C2" and "V2" denote the second principal component and so on. The row labeled "Eigenvalue" reports the eigenvalues of the sample second moment matrix in descending order from left to right. The "Variance Prop." row displays the variance proportion explained by each principal component. This value is simply the ratio of each eigenvalue to the sum of all eigenvalues. The "Cumulative Prop." row displays the cumulative sum of the Variance Prop. row from left to right and is the variance proportion explained by principal components up to that

Table 10.4 PCA test results for EUR/GBP

	C1	C2	C3	C4	C5	C6	C7	C8	C9	C10	C11
Eigenvalue	3.69	1.37	1.09	1.01	0.98	0.95	0.84	0.61	0.21	0.20	0.05
Var. Prop.	0.34	0.12	0.10	0.09	0.09	0.09	0.08	0.06	0.02	0.02	0.01
Cum. Prop.	0.34	0.46	0.56	0.65	0.74	0.83	0.90	0.96	0.98	0.99	1.00

	V1	V2	V3	V4	V5	V6	V7	V8	V9	V10	V11
EUR/USD	0.03	−0.20	−0.04	0.68	−0.64	0.03	−0.29	−0.01	0.02	−0.01	−0.01
GACGB10	0.07	0.20	−0.70	0.02	−0.30	0.02	−0.61	0.07	0.01	−0.04	0.00
GJGB10	0.05	0.26	−0.61	−0.11	−0.43	−0.09	0.58	0.15	0.01	0.01	0.00
IBEX	0.47	0.08	0.09	−0.04	−0.05	−0.02	−0.02	0.07	−0.61	−0.61	−0.11
JPY/USD	−0.08	0.66	0.14	−0.11	−0.25	−0.07	−0.18	−0.66	0.02	−.01	0.00
LMNIDY	0.08	0.27	−0.02	0.48	0.33	0.69	0.30	−0.10	0.00	−0.02	0.00
CAC	0.49	0.06	0.06	−0.03	−0.02	−0.01	−0.05	0.02	−0.09	0.60	−0.61
MIB30	0.47	−0.05	0.10	−0.05	−0.02	−0.03	−0.01	0.06	0.79	−0.37	−0.08
PLAT	−0.02	0.51	0.29	−0.08	−0.18	0.17	−0.21	0.71	0.01	0.04	0.00
CHF/USD	−0.20	0.50	0.29	−0.08	−0.18	0.17	−0.21	0.71	0.01	0.04	0.00
DJST	0.502	0.06	0.07	−0.02	−0.02	−0.01	−0.04	0.03	−0.08	0.35	0.78

Table 10.5 PCA test results for EUR/USD

	C1	C2	C3	C4	C5	C6	C7
Eigenvalue	1.76	1.07	0.99	0.97	0.91	0.75	0.55
Var. Prop.	0.25	0.15	0.14	0.14	0.13	0.11	0.08
Cum. Prop.	0.25	0.40	0.55	0.68	0.81	0.92	1.00
	V1	V2	V3	V4	V5	V6	V7
LMSNDY	0.49	−0.02	−0.15	0.21	0.05	0.80	−0.22
LMZSDY	0.56	−0.08	−0.07	0.11	−0.01	−0.55	−0.60
KC1	0.03	−0.44	−0.67	0.59	0.01	0.01	0.04
LMCADY	0.60	−0.01	−0.02	0.10	0.02	−0.19	0.77
SPX	0.21	−0.23	0.60	−0.51	−0.52	0.13	−0.04
AUDUSD	0.18	0.52	0.16	−0.56	0.59	0.03	−0.06
CO1	0.05	0.69	−0.37	−0.10	−0.61	−0.01	−0.01

Table 10.6 PCA test results for EUR/JPY

	C1	C2	C3	C4	C5	C6	C7	C8	C9	C10
Eigenvalue	2.30	1.22	1.05	1.02	1.01	0.96	0.79	0.67	0.65	0.33
Var. Prop.	0.23	0.12	0.11	0.10	0.10	0.10	0.08	0.07	0.07	0.03
Cum. Prop.	0.23	0.35	0.46	0.56	0.66	0.76	0.84	0.90	0.97	1.00
	V1	V2	V3	V4	V5	V6	V7	V8	V9	V10
GACGB10	0.35	0.45	0.12	−0.26	−0.14	0.08	−0.03	−0.09	−0.75	−0.06
GDBR10	0.55	−0.08	−0.15	0.25	−0.00	0.04	−0.09	−0.06	0.17	−0.75
GJGB10	0.16	−0.01	−0.06	−0.49	−0.18	−0.81	−0.13	0.04	0.17	−0.02
GSWISS10	0.50	−0.04	−0.17	0.13	0.05	0.03	−0.16	−0.58	0.16	0.55
GT10	0.45	−0.10	0.10	0.31	−0.12	−0.06	0.01	0.74	−0.05	0.35
IBEX	0.25	−0.41	0.17	−0.31	0.24	0.03	0.76	−0.07	−0.05	0.00
SILV	0.00	−0.12	−0.30	−0.24	−0.82	0.35	0.17	0.01	0.14	0.03
AU0003M	0.16	0.60	0.34	−0.27	0.10	0.25	0.10	0.11	0.57	0.02
BPSW10-GUKG10	−0.06	−0.02	0.70	0.39	−0.44	−0.23	0.13	−0.30	0.03	−0.05
EURCHF	0.08	−0.48	0.44	−0.38	0.03	0.30	−0.57	0.03	−0.01	−0.02

order. The second part of the output table displays the eigenvectors corresponding to each eigenvalue. The first principal component is computed as a linear combination of the series in the group with weights given by the first eigenvector. The second principal component is the linear combination with weights given by the second eigenvector and so on. Per cumulative R^2, which we required to be not lower than 0.99, significant multicollinearity could not be detected for any dependent variable. Consequently, the datasets were not reduced any further and all variables were kept.

10.2 SVM Model

We use the C-support vector classification (C-SVC) algorithm as described in [83, 405] and implemented in R packages "e1071" [70] and "kernlab" [220]:

Problem 10.1. Given training vectors $x_i \in R^n$, $i = 1,\dots,l$, in two classes, and a vector $y \in R^l$ such that $y \in \{+1,-1\}$, C-SVC solves the following problem:

$$\min_{w,b,\zeta} \frac{1}{2} w^T w + C \sum_{i=1}^{l} \zeta_i$$

w.r.t.

$$y_i \left(w^T \phi(x_i) + b \right) \geq 1 - \zeta_i \tag{10.3}$$

$$\zeta_i \geq 0, i = 1, \dots, l$$

Its dual representation is

$$\min_{\alpha} \frac{1}{2} \alpha^T Q \alpha - e^T \alpha$$

w.r.t. $\tag{10.4}$

$$0 \leq \alpha_i \leq C, \ i = 1,\dots,l$$

$$y^T \alpha = 0$$

where e is the vector of all ones, C is the upper bound, Q is an $l \times l$ positive semidefinite matrix, $Q_{ij} \equiv y_i y_j K(x_i,x_j)$, and $K(x_i,x_j) \equiv \phi(x_i)^T \phi(x_j)$ is the kernel, which maps training vectors x_i into a higher dimensional, inner product feature space by the function ϕ. The decision function is

$$f(x) = \text{sgn} \left(\sum_{i=1}^{l} y_i y_j K(x_i,x) + b \right) \tag{10.5}$$

10.3 Sequential Minimization Optimization (SMO) Algorithm

Training of a SVM requires the solution of a quadratic programing problem (QP), i.e., maximizing a convex quadratic form subject to linear constraints. Such convex quadratic programs have no local maxima and their solution can always be found analytically. Furthermore, the dual representation of the problem showed how training can be successfully affected even in very high dimensional feature spaces. The problem of minimizing differentiable functions of many variables has been widely studied, especially in the convex case, and many standard approaches can be directly applied to SVM training.[1] For large learning problems with many training examples, however, the constrained quadratic optimization approach to solving SVMs quickly becomes intractable in terms of time and memory requirements. For instance, a training set of 50,000 examples will yield a Q-matrix with 2.5 billion elements, which cannot fit easily into the memory of a standard computer. Consequently, traditional optimization algorithms, such as Newton or Quasi-Newton, cannot be directly applied. Several researchers have proposed decomposition methods to solve this optimization problem [55, 207, 222, 274, 317, 326, 404].

Vapnik et al. describe a "chunking" method [55] using the fact that the solution of a QP problem is the same if we remove the rows and columns of the matrix Q that correspond to zero Lagrange multipliers. Thus, a large QP problem can be decomposed into a series of smaller QP subproblems in which all of the nonzero Lagrange multipliers are identified and all zero Lagrange multipliers are discarded. After all the nonzero Lagrange multipliers in Q have been identified, the last step then solves the remaining QP problem. While chunking reduces the size of the matrix from N^2 (where N is the number of training examples) to the number of nonzero Lagrange multipliers, it still does not handle large scale training problems since the matrix still may not fit in the memory. Osuna [317] proposed a new decomposition algorithm for solving the SVM QP problem by showing that a large QP problem can be broken into a series of subproblems by maintaining a small working set. At each iteration step one or more examples that violate the KKT conditions are added to the smaller QP. Osuna's decomposition algorithm uses a constant size matrix for every subproblem and adds/subtracts one example at every step. Since the working set is usually small, this method does not have memory problems. However, a numerical QP solver is still required which raises numerical precision issues. Joachims [207] improves Osuna's methods with a strategy to select good working sets.

The algorithm that is opted for in the underlying context, however, is Platt's sequential minimization optimization (SMO) algorithm [326]. SMO is based on the idea that the quadratic programing problems can be broken up into a series of smallest possible QP problems which can be solved analytically. SMO solves the smallest possible optimization problem at every step, by

- Creating working sets of size 2 (involving two Lagrange multipliers) with a set of heuristics
- Jointly optimizing the two Lagrange multipliers

The main advantage is that the solution for the multipliers at each step is analytic and no QP solver is used. By avoiding the large matrix computation, SMO can handle very large training sets in between linear and quadratic time with a linear

---

[1] The first iterative algorithm for separating points from two populations by means of a hyperplane is the perceptron algorithm proposed by [345]. This algorithm starts with an initial weight vector $w_0 = 0$ and adapts it each time a training point is misclassified by the current weights. In its primal form the algorithm updates the weight vector and bias directly whereas in its dual form the final hypothesis is a linear combination of the training points. This procedure is guaranteed to converge provided there exists a hyperplane (i.e., there is a hyperplane whose geometric margin is positive) that correctly classifies the training data. In this case the data are said to be linearly separable. If no such hyperplane exists the data are said to be nonseparable.

amount of memory in the training set size. For comparison, standard approaches such as chunking and Osuna's algorithm can be on the order of cubic time. SMO can therefore be seen as a special case of the Osuna decomposition algorithm, which at the same time improves the latter method. Comparative testing against other algorithms, done by Platt, has shown that SMO is often much faster and has better scaling properties [326].

10.4 Kernel Selection

Ever since the introduction of the SVM algorithm, the question of choosing the kernel has been considered as very important. This is largely due to the effect that the performance highly depends on data preprocessing and less on the linear classification algorithm to be used. How to efficiently find out which kernel is optimal for a given learning task is still a rather unexplored problem and subject to intense current research. One approach is to use cross validation to select the parameters of the kernels and SVMs [104, 291] with varying degrees of success. The notion of Kernel target alignment [86, 87] uses the objective function of kernels spanned by the eigenvectors of the kernel matrix of the combined training and test data. The semidefinite programing (SDP) approach [241] uses a more general class of kernels, namely a linear combination of positive semidefinite matrices. They minimize the margin of the resulting SVM using a SDP for kernel matrices with constant trace. Similar to this, [56] further restrict the class of kernels to the convex hull of the kernel matrices normalized by their trace. This restriction, along with minimization of the complexity class of the kernel, allows them to perform gradient descent to find the optimum kernel. Using the idea of boosting, [84] optimize where are the weights used in the boosting algorithm. The class of base kernels is obtained from the normalized solution of the generalized eigenvector problem. Another possibility is to learn the kernel using Bayesian methods by defining a suitable prior, and learning the hyperparameters by optimizing the marginal likelihood [417, 418]. As an example of this, when other information is available, an auxiliary matrix can be used with the EM algorithm for learning the kernel [397]. Furthermore, [314] recently proposed a statistical inference framework for making kernels that are adaptive and allow independent scales for each dimension. This can be achieved by defining a reproducing Kernel Hilbert space (RKHS) on the space of kernels itself in order to let hyperkernels parameterize the kernels for each dimension.

We choose to take a pragmatic approach by comparing a range of kernels with regards to their effect on SVM performance. The argumentation is that even if a strong theoretical method for selecting a kernel is developed, unless this can be validated using independent test sets on a large number of problems, methods such as bootstrapping and cross validation will remain the preferred method for kernel selection. In addition, with the inclusion of many mappings within one framework it is easier to make a comparison. Standard kernels, which satisfy Mercer's conditions and whose performance is compared within the underlying classification task, include the following:

$$\text{Linear} : k(x,x') = \langle x,x' \rangle$$

$$\text{Polynomial} : k(x,x') = (scale \cdot \langle x,x' \rangle + offset)^{degree}$$

$$\text{Laplace} : k(x,x') = \exp(-\sigma \parallel x - x' \parallel)$$

$$\text{Gaussian Radial Basis Function} : k(x,x') = \exp(-\sigma \parallel x - x' \parallel^2) \quad (10.6)$$

$$\text{Hyperbolic} : k(x,x') = \tanh(scale \cdot \langle x,x' \rangle + offset)$$

$$\text{Bessel} : k(x,x') = \frac{\text{Bessel}_{v+1}^n(\sigma \parallel x - x' \parallel)}{(\parallel x - x' \parallel)^{-n(v+1)}}$$

The estimated parameters are provided in Table 10.7. In addition, we verify the use of the exotic p-Gaussian kernel

$$K(x_i,x_j) = \exp\left(-\frac{d(x_i,x)^p}{\sigma^p}\right) \quad (10.7)$$

To our knowledge, the p-Gaussian has hardly been tested yet on real datasets, but in theory, has interesting properties. Compared to the widely used RBF kernels, p-Gaussians include a supplementary degree of freedom in order to better adapt to the distribution of data in high-dimensional spaces [131]. The p-Gaussian is therefore determined by two parameters $p, \sigma \in R$ which define the Euclidean distance

$$d(x_i,x) = \left(\sum_{i=1}^n \|x_i - x\|^2\right)^{1/2} \quad (10.8)$$

between data points. The two parameters depend on the specific input set for each exchange rate return time series and are calculated as proposed in [131]:

$$p = \frac{\ln\left(\frac{\ln(0.05)}{\ln(0.95)}\right)}{\ln\left(\frac{dF}{dN}\right)}; \quad \sigma = \frac{d_F}{(-\ln(0.05))^{1/p}} = \frac{d_N}{(-\ln(0.95))^{1/p}} \quad (10.9)$$

In the case of EUR/USD, for instance, we are considering 1,737 eight-dimensional objects from which we calculate the kernel matrix. We consider 1,737 eight-dimensional objects because at each $1 \leq t \leq 1,737$, the similarity among the eight variables, as expressed by their Euclidean distance, has to be determined. The elements of the kernel matrix are then given by the single Euclidean distances of one variable to the remaining variables of the same system according to (10.8). For this reason $1,737 \times 1,737$ Euclidean distances, based on eight coordinates, have to be calculated in order to compute the 5% (d_N) and 95% (d_F) percentiles in that distribution which are needed to estimate p and s. The estimated parameters for all exchange rates are provided in Table 10.8.

Table 10.7 SVM parameter estimates with standard kernel functions

	Linear			Polynomial		
	EUR/GBP	EUR/JPY	EUR/USD	EUR/GBP	EUR/JPY	EUR/USD
SVs	1,663	1,684	1,650	1,646	1,611	1,629
Cost	1	1	1	1	1	1
Degree	–	–	–	–	–	–
Scale	–	–	–	–	–	–
Gamma	0.09	0.10	0.14	0.09	0.10	0.14
Coef 0	–	–	–	–	–	–
Offset	–	–	–	–	–	–
σ	–	–	–	–	–	–

	Laplace			Gaussian		
	EUR/GBP	EUR/JPY	EUR/USD	EUR/GBP	EUR/JPY	EUR/USD
SVs	1,688	1,709	1,685	1,636	1,662	1,630
Cost	1	1	1	1	1	1
Degree	–	–	–	–	–	–
Scale	–	–	–	–	–	–
Gamma	–	–	–	0.09	0.10	0.14
Coef 0	–	–	–	–	–	–
Offset	–	–	–	–	–	–
σ	0.04	0.05	0.06	–	–	–

	Hyperbolic			Bessel		
	EUR/GBP	EUR/JPY	EUR/USD	EUR/GBP	EUR/JPY	EUR/USD
SVs	878	912	930	1,182	1,268	1,472
Cost	1	1	1	1	1	1
Order	–	–	–	1	1	1
Degree	–	–	–	1	1	1
Scale	1	1	1	–	–	–
Gamma	–	–	–	–	–	–
Coef 0	–	–	–	–	–	–
Offset	1	1	1	–	–	–
σ	–	–	–	1	1	1

Table 10.8 p-Gaussian model estimates

	EUR/GBP	EUR/JPY	EUR/USD
SVs	1571	1584	1571
Cost	1	1	1
d_F	0.1207	0.1781	0.1039
d_N	0.0307	0.0338	0.0254
p	2.9678	2.4482	2.8854
σ	0.0834	0.1138	0.0711

10.5 Cross Validation

The development of multivariate models generally requires checking their validity. Whereas the reliability of linear models is normally expressed by theoretically justified metrics, such as R^2, it is more difficult to examine the reliability of nonlinear methods, such as supervised learning methods. Cross validation is a common technique for estimating the quality (accuracy) of a classifier induced by supervised learning algorithms. In addition, cross validation may be used for reasons of model selection, i.e., choosing a classifier from a given set or combination of classifiers ([420]). In k-fold cross validation, the dataset D is randomly split into k mutually exclusive subsets of approximately equal size. Each time $t \in \{1,\ldots,k\}$, the learning algorithm is trained on $D \backslash D_t$ and tested k times on D_t. The cross validation accuracy is the overall number of correct classifications, divided by the number of instances in the dataset, i.e., the average accuracy rate of the k single runs. Formally, if $D_{(i)}$ denotes the test set that includes instance $x_i = (x_i, y_i)$, then the cross validation accuracy is given by

$$acc_{CV} = \frac{1}{n} \sum_{(x_i,y_i) \in D} \delta \left(\Gamma(D \backslash D_{(i)}, x_i), y_i \right) \qquad (10.10)$$

Complete cross validation is the average of all possibilities for choosing m/k instances out of m. However, this can be a very time consuming process. In our case, the underlying in-sample dataset consisting of 1,737 instances is divided into $k = 5$, $k = 10$, $k = 15$, and $k = 20$ subsets. Now k test runs are initiated where $(k-1)/k$ instances are considered for training and $1/k$ are considered for testing. This process is being repeated for all subsets until every subset has been considered once for testing. Tables 10.9 and 10.10 display the computing times required for training above specified SVMs per currency pair. Table 10.9 describes the runtime behavior of SVMs with standard kernels, whereas Table 10.10 separately summarizes the computational expenses obtained when training the SVMs with a p-Gaussian.

Obviously, the more subsets we choose to validate a model, the more increases training time. Moreover, if we rank the SVM models from lowest to highest with regards to their required running-times, we end up at the following ordering: Linear $<$ Hyperbolic $<$ Gaussian RBF $<$ Polynomial $<$ Laplace $<$ Bessel $<$ p-Gaussian. This can be explained by the input lengths of the different kernel functions (see (10.6)). The most basic kernel function requiring the least arithmetic operations is the linear kernel. The most sophisticated kernel, requiring by far the most the computing time, is the p-Gaussian (compare Tables 10.9 and 10.10). This can be attributed to a noninteger p in the exponent of the Euclidean distance matrix. It should also be mentioned that due to the random nature of dividing the data into subsets, the single accuracy rates change each time. This leads to a change of the total accuracy rate. If the change is significant, which could not be confirmed in our analyses, it is often recommended to change or optimize the SVM parameter.

Table 10.9 Runtime behavior of standard kernels

	Linear			Polynomial		
	EUR/GBP	EUR/JPY	EUR/USD	EUR/GBP	EUR/JPY	EUR/USD
No CV	00:02	00:01	00:01	00:02	00:01	00:01
5-fold CV	00:05	00:04	00:03	00:05	00:04	00:05
10-fold CV	00:09	00:07	00:06	00:10	00:11	00:10
15-fold CV	00:12	00:11	00:09	00:15	00:15	00:15
20-fold CV	00:16	00:15	00:12	00:20	00:21	00:21
	Laplace			Gaussian		
	EUR/GBP	EUR/JPY	EUR/USD	EUR/GBP	EUR/JPY	EUR/USD
No CV	00:05	00:05	00:05	00:02	00:02	00:01
5-fold CV	00:11	00:11	00:10	00:05	00:05	00:05
10-fold CV	00:17	00:17	00:15	00:10	00:10	00:09
15-fold CV	00:22	00:22	00:20	00:14	00:14	00:14
20-fold CV	00:27	00:27	00:26	00:18	00:18	00:17
	Hyperbolic			Bessel		
	EUR/GBP	EUR/JPY	EUR/USD	EUR/GBP	EUR/JPY	EUR/USD
No CV	00:02	00:02	00:02	00:08	00:08	00:10
5-fold CV	00:04	00:04	00:04	00:17	00:17	00:20
10-fold CV	00:07	00:07	00:07	00:22	00:23	00:26
15-fold CV	00:11	00:10	00:10	00:27	00:29	00:32
20-fold CV	00:16	00:14	00:13	00:32	00:35	00:38

Table 10.10 Runtime behavior of p-Gaussian

	p-Gaussian		
	EUR/GBP	EUR/JPY	EUR/USD
No CV	02:02	02:01	01:55
5-fold CV	05:55	05:55	05:42
10-fold CV	10:52	10:50	10:39
15-fold CV	18:32	18:27	18:25
20-fold CV	28:15	28:06	27:52

10.6 Benchmark Models

Letting y_t represent the exchange rate at time t, we forecasted the variable

$$\text{sgn}(\Delta y_{t+h}) = \text{sgn}(y_{t+h} - y_t) \tag{10.11}$$

where $h = 1$ for a one-period forecast with daily data. Even small tremors were interpreted as real changes of directions.

Kernel-dependent SVM forecasts were compared to the forecasts of two kinds of benchmark models: the naive model and the ARMA(p,q) model. The naive strategy assumes that the most recent period change is the best predictor of the future, i.e.,

$$\text{sgn}(\Delta \hat{y}_{t+1}) = \text{sgn}(\Delta y_t) \tag{10.12}$$

where Δy_t is the actual rate of return at period t and $\Delta \hat{y}_{t+1}$ is the predicted rate of return for the next period.

The ARMA(p,q) model as described in Sect. 8.2.3 is the most general family of models for representing stationary processes and therefore represents a suitable benchmark. We use the model estimates as given by (8.12)–(8.14)

- $\Delta y_t = -0.0526 \Delta y_{t-1} - 0.0562 \Delta y_{t-3} + \varepsilon_t$ for the EUR/GBP series
- $\Delta y_t = 0.0287 \Delta y_{t-1} + \varepsilon_t$ for the EUR/JPY series
- $\Delta y_t = -0.5959 \Delta y_{t-1} + 0.5323 \varepsilon_{t-1} + \varepsilon_t$ for the EUR/USD series

The EUR/GBP and EUR/JPY models are AR(p) models. Dynamic forecasting with AR(p) models is done by generating the one-step ahead forecast and then using this for a two-step-ahead forecast and so on. That is, the optimal one-step-ahead prediction at time T is the conditional expectation of y_{T+1} given $\{y_T, y_{T-1}, y_{T-2}, \ldots\}$:

$$\hat{y}_{T+1} - \hat{c} = \hat{\alpha}_1 (y_T - \hat{c}) + \hat{\alpha}_2 (y_{T-1} - \hat{c}) + \ldots + \hat{\alpha}_p (y_{T-p+1} - \hat{c}) \tag{10.13}$$

and the two-step prediction is:

$$\hat{y}_{T+2} - \hat{c} = \hat{\alpha}_1 (y_{T+1} - \hat{c}) + \hat{\alpha}_2 (y_T - \hat{c}) + \ldots + \hat{\alpha}_p (y_{T-p+2} - \hat{c}) \tag{10.14}$$

and so on. Since in our case $\hat{c} = 0$, the s-step-ahead forecast for the EUR/GBP model is given by

$$\hat{y}_{T+s} = \hat{\alpha}_1 y_{T+s-1} + \hat{\alpha}_3 y_{T+s-3} \tag{10.15}$$

and the s-step-ahead forecast for the EUR/JPY model is

$$\hat{y}_{T+s} = \hat{\alpha}_1 y_{T+s-1} \tag{10.16}$$

For a general ARMA(p,q) model the s-step-ahead predictions are

$$\begin{aligned}
\hat{y}_{T+s} - \hat{c} &= \hat{\alpha}_1 (\hat{y}_{T+s-1} - \hat{c}) \\
&+ \hat{\alpha}_2 (\hat{y}_{T+s-2} - \hat{c}) + \ldots + \hat{\alpha}_p (\hat{y}_{T+s-p} - \hat{c}) \\
&+ \hat{\beta}_s \varepsilon_T + \hat{\beta}_{s+1} \varepsilon_{T-1} + \ldots + \hat{\beta}_q \varepsilon_{T-q-s}
\end{aligned} \tag{10.17}$$

for $s \leq q$. For $s > q$ only the AR part determines the forecasts. The EUR/USD model is an ARMA(1,1) model with $\hat{c} = 0$ whose s-step-ahead predictions are therefore

$$\hat{y}_{t+s} = \hat{\alpha}_1 y_{T+s-1} + \hat{\beta}_s + \varepsilon_t \tag{10.18}$$

Note that the GARCH models, as estimated in Sect. 8.2.4 are not useful for serving as benchmark models within the realm of tackling the BCP. The purpose of this family of models is to describe the development of the second statistical moment, whereas the task here is to predict directional movements, i.e., ups and downs of the exchange rate.

10.7 Evaluation Procedure

The evaluation procedure is twofold. Out-of-sample forecasts are evaluated both statistically via confusion matrices and practically via trading simulations.

10.7.1 Statistical Evaluation

Generally, a predictive test is a single evaluation of the model performance based on comparison of actual data with the values predicted by the model. We consider the BCP where, formally, each instance I is mapped to one element of the set $\{-1, +1\}$ of positive and negative class labels by a classification model. Given a classification model and an instance, there are four possible outcomes. If the instance is positive and it is classified as positive, it is counted as a true positive. If it is classified as negative, it is counted as a false negative. If the instance is negative and it is classified as negative, it is counted as a true negative. If it is classified as positive, it is counted as a false positive. Given a classification model and a set of instances (the test set), a two-by-two confusion matrix (also called a contingency table), as is illustrated in Fig. 10.1, can be constructed representing the dispositions of the set of instances. This confusion matrix forms the basis for many statistical performance metrics (see for instance [127], pp. 2–4). Since we are equally interested in predicting ups and downs, the accuracy rate

$$Accuracy = \frac{TP + TN}{P + N} \tag{10.19}$$

defined as the sum of true positives TP and true negatives TN divided by the sum of the total amount of positive observations P and negative observations N is the right statistical performance measure to apply.

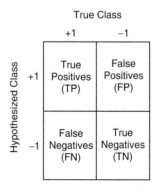

Fig. 10.1 Confusion matrix

10.7.2 Operational Evaluation

However, potential evidence for the predictability of exchange rate returns does not necessarily imply market inefficiency. It is likely that the small average excess returns would not generate net gains when employed in a trading strategy once transaction costs have been taken into account. This idea has been referred to as the concept of speculative efficiency which implies the absence of unexploited speculative profits. Therefore, under this definition of market efficiency, these markets would not be classified as inefficient. Practical or operational evaluation methods focus on the context in which the prediction is used by imposing a metric on prediction results which permits testing above hypothesis. More generally, when predictions are used for trading or hedging purposes, the performance of a trading or hedging metric provides a measure of the model's success. We formulate a trading simulation and calculate its profitability in simulated trading. First of all, return predictions y_t were translated into positions. Next, a decision framework was established that tells us when the underlying asset is bought or sold depending on the level of the price forecast.

$$ I_t = \begin{cases} 1 & : \quad \hat{y}_t < y_{t-1} - \tau \\ -1 & : \quad \hat{y}_t > y_{t-1} + \tau \\ 0 & : \quad \text{otherwise} \end{cases} \tag{10.20} $$

If the model predicts a negative return, the trader finds himself in a long position ($I_t = 1$). A long position is characterized by a "sell" signal concerning the base currency (EUR) as per today. The intuition is as follows: if the exchange rate is believed to go down tomorrow, then one should sell the base currency today, since 1 EUR is worth more in terms of foreign currency today than it will be tomorrow. Otherwise, if the model predicts a positive return, the trader finds himself in a short position ($I_t = -1$). A short position is characterized by a "buy" signal concerning the base currency (EUR) as per today. The intuition is as follows: if the exchange rate is believed to go up tomorrow, then one should buy the base currency today, since it is cheaper in terms of foreign currency than it will be tomorrow.

The parameter t describes a subjective threshold, which can be added to the last price change leading to different signals. This may be useful if one intends to follow a longer run strategy where t guarantees that the base currency will not be bought/sold above/below a certain level. We decided to set $t = 0$.

For measuring prediction performance on the operational level, the following nine metrics were chosen: cumulated P/L, shape ratio as the quotient of annualized P/L and annualized volatility, maximum daily profit, maximum daily loss, maximum drawdown, value-at-risk with 95% confidence, average gain/loss ratio, and trader's advantage. A formal definition for each of these metrics is given in Table 10.11. Note that every metric is somewhat dependent on the gain or loss

$$ \pi_t = I_{t-1}(y_t - y_{t-1}) \tag{10.21} $$

on the position at time t but makes a unique statement. Accounting for transaction costs (TC) is important in order to assess trading performance in a realistic way.

Table 10.11 Operational performance measures

Measure	Definition
Cumulated profit and loss	$PL_T^C = \sum_{t=1}^{T} \pi_t$
Sharpe ratio	$SR = \frac{PL_T^A}{\sigma^A}$, with $PL_T^A = 252 \times \frac{1}{T} \sum_{t=1}^{T} \pi_t$ and
	$\sigma_T^A = \sqrt{252} \times \sqrt{\frac{1}{T-1} \times \sum_{t=1}^{T} (\pi_t - \bar{\pi})^2}$
Maximum daily profit	$\max(\pi_1, \pi_2, \ldots, \pi_T)$
Maximum daily loss	$\min(\pi_1, \pi_2, \ldots, \pi_T)$
Maximum drawdown	$MD = \min(PL_t^C - \max_{i=1,\ldots,t}(PL_i^C))$
Value-at-Risk	$VaR = \mu - Q(\pi, 0.05)$, $\mu = 0$
Net cumulated profit and loss	$NPL_T^C = \sum_{t=1}^{T}(\pi_t - I_t \times TC)$, where $I_t = 1$ if $\pi_{t-1} \times \pi_t < 0$, else $I_t = 0$
Average gain/loss	$\frac{AG}{AL} = \frac{(\text{Sum of all } \pi_t > 0)/\# \text{ up}}{(\text{Sum of all } \pi_t < 0)/\# \text{ down}}$
Trader's advantage	$TA = 0.5 \times \left(1 + \left(\frac{(WT \times AG) + (LT \times AL)}{\sqrt{(WT \times AG^2) + (LT \times AL^2)}}\right)\right)$ with $WT :=$ number of winning trades, $LT :=$ number of losing trades, $AG :=$ average gain in up periods, and $AL :=$ average loss in down periods

Between market-makers an average cost of three pips (0.0003) per trade for a tradable amount of typically 10–20 million EUR is considered as a reasonable guess and thus incorporated in net cumulated profit.

10.8 Numerical Results and Discussion

In order to compare forecasts for the same series across different models, accuracy rates for the out-of-sample period are depicted by bar charts as shown in Figs. 10.2–10.4.

One immediate drawback may be the obtained accuracy, which is only 5% better for the best SVM models compared to the simple naive predictor. However, it should be stressed that the tested datasets are very small due to the short history of the Euro. As argued by [246], there are important data mining applications where the data is scarce and more research is needed towards methods that can deal with such datasets. This work backs this claim.

Furthermore, Part II has shown that forecasting exchange rate directions is a very difficult task. Foreign exchange markets are highly liquid and considered as very efficient, and it is difficult to gain superior information out of publicly available data. Consequently, if SVM accuracy rates outperform those of nave or random strategies, the SVM technique can be generally justified to predict exchange rate return directions. This is the case for all three exchange rates. In particular, we find that hyperbolic SVMs deliver superior performance for out-of-sample prediction across all three currency pairs. In the case of EUR/GBP, the Laplace SVM performs equally

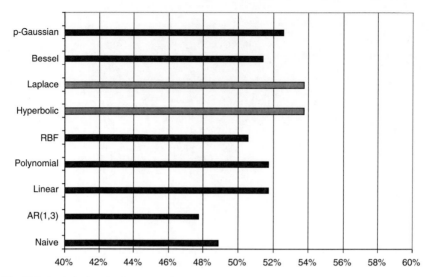

Fig. 10.2 Statistical evaluation: EUR/GBP

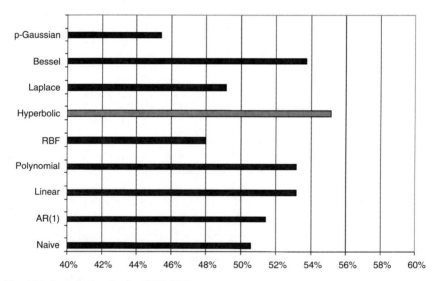

Fig. 10.3 Statistical evaluation: EUR/JPY

well as the hyperbolic SVM. Other models are out-performed by the hyperbolic kernel SVM more clearly in the cases of EUR/JPY and EUR/USD. This observation
makes hyperbolic kernels promising candidates to map all sorts of financial market
return data into high dimensional feature spaces.

Tables 10.12–10.14 give the results of the trading simulation. The depicted values simply represent an aggregation of the EUR returns bought or sold with foreign

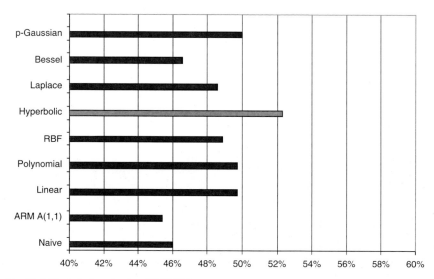

Fig. 10.4 Statistical evaluation: EUR/USD

Table 10.12 Operational performance: EUR/GBP

	Naive	ARMA	Linear	Polyn.	RBF	Hyper.	Laplace	Bessel	p-Gauss.
PL_C^T	−0.0075	−0.0911	−0.0936	−0.0936	−0.0390	0.1036	0.0155	−0.0411	0.0596
SR	−0.0797	−0.9670	−0.9937	−0.9937	−0.4135	1.0994	0.1641	−0.4367	0.6323
max P	0.0149	0.0149	0.0168	0.0168	0.0168	0.0149	0.0168	0.0139	0.0123
max L	−0.0168	−0.0168	−0.0149	−0.0149	−0.0139	−0.0168	−0.0139	−0.0168	−0.0168
MD	−0.0381	−0.0381	−0.0362	−0.0362	−0.0350	−0.0381	−0.0351	−0.0356	−0.0381
VaR	−0.0070	−0.0077	−0.0075	−0.0075	−0.0073	−0.0070	−0.0069	−0.0074	−0.0069
NPL_C^T	−0.0612	−0.1955	−0.1275	−0.1275	−0.0903	−0.0559	−0.0196	−0.0921	0.0143
AG/AL	1.0581	0.9279	0.8037	0.8037	0.9171	1.0398	0.8993	0.8823	1.0189
TA	0.0000	0.4185	0.5300	0.5300	0.4872	0.4814	0.5899	0.3935	0.4351

Table 10.13 Operational performance: EUR/JPY

	Naive	ARMA	Linear	Polyn.	RBF	Hyper.	Laplace	Bessel	p-Gauss.
PL_C^T	0.0544	0.0525	−0.0948	−0.0948	−0.2191	−0.1387	−0.2867	−0.3114	−0.2498
SR	0.3868	0.3729	−0.6743	−0.6743	−1.5568	−0.9862	−2.0360	−2.2112	−1.746
max P	0.0219	0.0219	0.0207	0.0207	0.0207	0.0217	0.0207	0.0207	0.0205
max L	−0.0205	−0.0217	−0.0219	−0.0219	−0.0219	−0.0219	−0.0219	−0.0219	−0.0219
MD	−0.0853	−0.0866	−0.0648	−0.0648	−0.0867	−0.0620	−0.0867	−0.0648	−0.0867
VaR	−0.0100	−0.0097	−0.0109	−0.0109	−0.0111	−0.0108	−0.0113	−0.0115	−0.0113
NPL_C^T	0.0028	0.0522	−0.1527	−0.1527	−0.2761	−0.1984	−0.3446	−0.3618	0.3026
AG/AL	1.0411	1.0033	0.9000	0.9000	0.8828	0.8646	0.8332	0.8375	0.8218
TA	0.0000	1.0000	0.4301	0.4301	0.4325	0.4365	0.4115	0.4035	0.4014

Table 10.14 Operational performance: EUR/USD

	Naive	ARMA	Linear	Polyn.	RBF	Hyper.	Laplace	Bessel	p-Gauss.
PL_C^T	−0.1807	−0.1035	−0.1326	−0.1326	−0.093	0.0480	−0.1005	−0.1617	0.1018
SR	−1.2345	−0.7052	−0.9043	−0.9043	−0.0630	0.3252	−0.6851	−1.1037	0.6890
$\max P$	0.0196	0.0196	0.0167	0.0167	0.0196	0.0196	0.0189	0.0187	0.0189
$\max L$	−0.0189	−0.0187	−0.0196	−0.0196	−0.0187	−0.0189	−0.0196	−0.0196	−0.0196
MD	−0.0417	−0.0439	−0.0448	−0.0448	−0.0439	−0.0441	−0.0448	−0.0448	−0.0448
VaR	−0.0125	−0.0110	−0.0126	−0.0126	−0.0118	−0.0108	−0.0118	−0.0117	−0.0112
NPL_C^T	−0.2368	−0.2079	−0.1743	−0.1743	−0.0597	0.0000	0.1452	−0.2106	0.0511
AG/AL	0.9471	1.0579	0.8812	0.8812	1.0362	0.9627	0.9457	0.9457	1.1087
TA	0.0000	0.4536	0.6253	0.6253	0.5683	0.5531	0.5838	0.4219	0.4992

currency independent of the notional amount. The following conclusions can be drawn. The hypothesis of speculative efficiency may be rejected if $NPL_C^T > 0$ holds for a particular strategy, i.e., if the cumulated profit after transaction costs over the considered trading period is positive. Consider Table 10.12. It is interesting to observe that only SVMs equipped with hyperbolic, Laplace, and p-Gaussian kernels are able to achieve a positive net cumulated profit.

However, if we look at the results for EUR/JPY as given in line 7 of Table 10.13, one may have serious concerns on whether SVMs are able to make speculative profits. In fact, all trading strategies result in a negative net cumulated profit and the nave strategy is the only one generating a positive net cumulated profit. For EUR/USD, however, the p-Gaussian SVM is able to reject the hypothesis of speculative efficiency by generating positive net cumulated profit. To summarize, we can reject the speculative efficiency hypothesis since for all three exchange rates speculative profits can be made. The SVM approach has only merit for EUR/GBP and EUR/USD. Thus, in the spirit of computational complexity, the superior models seem to be acceptable approximations to a hypothetical optimal forecasting technique that would have total knowledge about new information.

A different way of analyzing the results could be as follows: let the best forecasting technique be the one that does not only satisfy one criterion best, but that is superior with regards to multiple trading metrics. Dominant strategies are represented by the maximum value(s) in each row. Operational evaluation results confirm statistical ones in the case of EUR/GBP. Both the hyperbolic and the Laplace SVM give best results along with the RBF SVM. For EUR/JPY and EUR/USD the results differ. The statistical superiority of hyperbolic SVMs cannot be confirmed on an operational level which is contradictory to the EUR/JPY and EUR/USD operational results at first glance. The reason for this phenomenon stems from the fact that operational evaluation techniques do not only measure the number of correctly predicted exchange rate ups and downs, they also include the magnitude of returns. Consequently, if local extremes can be exploited, forecasting methods with less statistical performance may yield higher profits than methods with greater statistical performance. Thus, in the case of EUR/USD, the trader would have been better off applying a p-Gaussian SVM in order to maximize profit. Regarding EUR/JPY, we find that no single model is able to outperform the nave strategy. The hyperbolic SVM, however, still dominates two performance measures. We also see that p-Gaussian SVMs perform reasonably well in predicting EUR/GBP and EUR/USD return directions but not EUR/JPY. For the EUR/GBP and EUR/USD currency pairs, p-Gaussian data representations in high dimensional space lead to better generalization than standard Gaussians due to an additional degree of freedom p.

To our knowledge, this is the first time, financial time series directions in general, and exchange rate directions in particular have been approached by SVMs. Further exploratory research therefore needs to be performed which could focus on SVM model improvements, for instance, examination of other sophisticated kernels, proper adjustment of kernel parameters and the development of data mining and optimization techniques for selecting the appropriate kernel. In light of this research, it would also be interesting to see if the dominance of hyperbolic SVMs can be confirmed in further empirical investigations on financial market return prediction. Moreover, if we consider the evidence for second moment nonlinear dependencies from Sect. 8.2.4, another interesting avenue of future research could be to use SVMs in order to predict financial market volatilities.

Part IV
Exchange Rate Hedging
in a Simulation/Optimization Framework

"Human life occurs only once, and the reason we cannot determine which of our decisions are good and which are bad is that in a given situation we can make only one decision, we are not granted a second, third, or fourth life in which to compare various decisions." (Milan Kundera)

Chapter 11
Introduction

International investing and trade has one unintended consequence: the creation of currency risk which may cause the local currency value of a firm's foreign receivables, liabilities or investments to fluctuate dramatically because of pure currency spot movements. In their corporate risk management survey, [364] stated that industrial corporations rank foreign exchange risks as the most costly risk with 93% of firms reporting some kind of foreign exchange exposure, and, on average, firms judging between a quarter and a third of their revenues, costs and cash flows as being exposed to movements in exchange rates.

Protection against foreign exchange risks may be achieved through internal or external hedging. Internal hedging refers to the exploitation of possibilities of changing the currency of cash outflows to better align them with inflows. Instruments include invoicing imports and exports in the home currency, contracting currency and foreign exchange clauses, speeding up or slowing down payments (leading or lagging), matching or netting claims, currency reserves, changing debt/claim structures, adjusting credit conditions and prices, etc. [109, 396]. This can be done by changing vendors, by relocating production facilities abroad or by foreign-denominated debt, that is, by funding itself in the foreign currency. Such hedges can act as substitutes to currency derivatives [154, 307, 331]. External hedging instruments, besides derivatives, include export factoring and forfaiting, international leasing, and foreign exchange insurance [109]. However, because hedging normally refers to the use of off-balance-sheet instruments, that is forward-based and option-based contracts in the case of exchange rate hedging, only such currency derivatives shall be considered in the following.

Academic literature on currency and commodity hedging has developed several theories to explain an individual's incentive to hedge. The characteristics of the oldest concept go back to the works of [190, 215, 226] and are as follows. First, hedging made the hedger's position certain. He sold his product forward and made delivery under the contract. The sole purpose for an individual to hedge was certainty, or in other words, reduction of income variability. Keynes [226] is explicit that hedging eliminates risk:

> If this [futures] price shows a profit on his costs of production, then he can go full steam ahead, selling his product forward and running no risk. [...]. If supply and demand are

C. Ullrich, *Forecasting and Hedging in the Foreign Exchange Markets,* Lecture Notes in Economics and Mathematical Systems 623, DOI: 10.1007/978-3-642-00495-7_11,
© Springer-Verlag Berlin Heidelberg 2009

balanced, the spot price must exceed the forward price by the amount which the producer is ready to sacrifice in order to *hedge* himself, i.e., to avoid the risk of price fluctuations during his production period.

Hicks [190], although taking over Keynes's theory speaks of the risk of a price change being *reduced* rather than avoided:

> The ordinary businessman only enters into a forward contract if by doing so he can *hedge* – that is to say the forward transaction lessens the riskiness of his position.

Kaldor [215] describes hedgers as

> [...] those who have certain commitments independent of any transactions in the forward market, [...] and who enter the forward market in order to reduce the risks arising out of these commitments.

He therefore speaks of hedging as reducing rather than eliminating risks. The second characteristic of the early concept of hedging is that a hedger does not act on his expectations, i.e., he always sold/bought forward irrespective of any expectations about the future price. Keynes, Hicks, and Kaldor do not say explicitly that hedgers have no expectations, but they do not give the expected price any role in the hedger's actions. Regarding currency risk, [322] advocated a fully hedged currency position on the basis of foreign currency risk not offering a commensurate return. In what they deem a "free lunch," they argue that as a result of its zero long-term expected return, currency risk can be removed without the portfolio suffering any reduction in long-term return. Therefore, many analysts incorrectly came to the conclusion that the availability of extreme liquidity (and hence low bid-ask spreads), a long term zero return and an apparent lack of predictive power of academic currency models meant that investors and corporations should naively remove currency risk by implementing passive hedges back into the base currency as one could reduce volatility without paying for it.

Froot et al. [144] took exactly the opposite approach and suggested that investors should do nothing and leave investments unhedged and unmanaged. He argued that over long investment horizons, real exchange rates revert back to their means according to the theory of PPP and investors should maintain an unhedged foreign currency position. He also concludes that even over shorter horizons, the small transaction costs and counterparty risks associated with maintaining a currency hedge add up over time and cause the optimal hedge ratio to decline as the investment timeframe increases. Still, [144] does acknowledge that real exchange rates may deviate from their theoretical fair value over shorter horizons and currency hedging in this context is beneficial in dampening volatility. One decade earlier, [105] challenged the Miller–Modigliani hypothesis – which states that the issue of hedging and corporate risk management, in general, is closely linked to the irrelevance of a firm's financial policy in complete and frictionless capital markets [298] – in its pure form and demonstrated that hedging can add value if markets are imperfect and somewhat disconnected from neoclassical parity conditions. In the 1990s, a more detailed theoretical discussion evolved about these market imperfections and their effect on corporate practices for hedging foreign exchange risk and other financial risks [144, 145, 307, 370]. Market imperfections that justify static risk management policies are:

- Information asymmetry [93, 94]: Management knows about the firm's exposure position much better than investors. Thus, the management of the firm, not its investors, should manage exchange exposure.
- Managerial interests [370]: Large portions of managers' wealth are related to the well-being of the firm in the form of income and, possibly, share ownership. Since managers are not fully diversified, they have an incentive to hedge the risks inherent in their position.
- Differential transaction costs: The firm is in a position to acquire low-cost hedges whereas transaction costs for individual investors can be substantial. In addition, the firm has access to hedging tools that are not available to private investors.
- Default costs [370]: If default costs are significant, corporate hedging would be justifiable because it will reduce the probability of default. Perception of a reduced default risk, in turn, can lead to a better credit rating and lower financing costs.
- Progressive corporate taxes [284, 370]: Under progressive corporate tax rates, stable before-tax earnings lead to lower corporate taxes than volatile earnings with the same average value. This happens, because under progressive tax rates, the firm pays more taxes in high-earning periods than it saves in low-earning periods.
- Financing investments [144]: Firms should hedge to ensure they always have sufficient cashflow to fund their planned investment program.

Apart from these reasons, recent empirical studies find that aspirations of risk management programs also instill a culture within an organization which improves risk-based decision making among employees [364]. However, risk management also incorporates several difficulties. First, it creates transaction costs associated with the purchase of derivatives and insurance contracts. Second, potential upside in hedged risks may be eliminated. Third, establishing a risk management function creates direct organizational costs such as salaries for staff and costs of IT systems. Fourth, risk management also creates additional risks, for example, the possibility of trading losses. Finally, risk management is inherently difficult itself and therefore hard to communicate to investors and board members.

The hedging policy to time currency and commodity markets with forward contracts has been referred to as *selective hedging* [376]. This technique supports the early ideas of [422] who argued that hedging is essentially not a risk reduction technique only, but speculation in the *basis* which is given by the difference between the spot and forward price. He makes clear that hedging in his sense contains a speculative element when he says ([422], p. 320):

> Such discretionary hedging involving a firm in the practice of both hedging and speculation, [...] seems to be especially prevalent among dealers and processors who handle commodities such as wool and coffee.

For his theoretical work on hedging, Working is considered as a pioneer for many researchers arguing that individuals hedge to maximize profits, taking into account expected changes in prices. The association in the literature of hedging with the motive of an uncertain profit is a reasonable consequence of the fact that hedging does not result in certainty. If hedging leaves risk, then expectations become relevant; and if a loss from hedging is expected it would be unwise to hedge. On the other hand, if a profit is expected it would appear advantageous to hedge. It is interesting to notice, that although financial markets are, per definition, not the core business of nonfinancial companies, they seem to be confident enough to take views on the future development of the exchange rate that will turn out to increase the firms' cashflows. Empirical evidence suggests that European firms are more inclined than US firms to accept open foreign exchange positions based on exchange rate forecasts [47]. In fact, selective hedging is also the approach that most German companies adopt [159]. In its corporate risk management survey, [364] have lately supported this result stating that

> although risk management of FX rates [...] is common, risks are rarely totally eliminated.

It must be concluded that firms implicitly reject the efficient market hypothesis in its semi-strong version, by forming different expectations than those conveyed by forward prices.[1] Under the efficient market view of forward prices, hedgers are unlikely to profit consistently from forward pricing strategies, so risk aversion becomes the primary motive for using forward markets. However, if hedgers have different expectations from those portrayed by forward prices, then their use of forward markets can be for the purpose of increasing profits as a result of those differing expectations, rather than managing risk. A large body of research has derived optimal hedging rules under this assumption [213, 242, 305, 321, 352]. However, one can also argue that financial markets in reality display a high degree of information efficiency because so many private and professional market participants are continuously striving to gain access to new and better information and to analyze the available information very carefully. This argument may be seen as a warning for selective hedgers. In order to achieve the goal of increasing cashflows, firms willingly accept the risk of currency losses due to the open positions. The speculative nature of the selective hedging strategy has been pointed out very sharply by [252], pp. 198–199:

> In fact, to the extent that it includes a speculative element by factoring possible gains into the hedging decision, [selective hedging] differs little from staking the assistant treasurer with a sum of money to be used to speculate on stock options, pork bellies or gold.

Stulz [376] stretches that selective hedging will increase shareholder value if managers have an informational advantage relative to other market participants. However, if managers believe they have informational advantages when they do not, selective hedging will merely result in an increase in the variability of cash flows that could potentially reduce shareholder value. In fact, [65] reports that economic gains to selective hedging are very small and less than for an alternative technical

---

[1] A strict interpretation of the efficiency hypothesis is not very plausible because in this case nobody would have an incentive to invest in the production or analysis of new information.

trading strategy. He also finds no evidence that selective hedging leads to superior operating or financial performance, nor is it associated with proxies for superior market information. These results emphasize that the main advantage of the selective hedging strategy, namely the granted flexibility for the decision-maker to actively react to market events, may inhibit a lot of risk. Finding an acceptable or possibly optimal balance between risk and reward is therefore relevant for any firm pursuing a selective strategy to hedging its currency risk.

In this context, the popular concept of maximizing income subject to a given level of risk [279], or alternatively, maximizing the level of expected utility derived from the activities being considered [411] becomes relevant.[2] The two approaches of hedging in order to reduce/eliminate risk [190, 215, 226] and hedging in order to maximize profit [422] become special cases of Markowitz's portfolio-theoretic approach when either risk or return is considered to be extremely important. While the mean, as the central moment of a probability distribution, is commonly accepted as a measure of (average) return, the adequacy of different kinds of risk measures has been and is still being discussed to an increasing extent [3, 4, 17, 177, 212, 279, 294, 315, 334, 339]. Mean-variance analysis is consistent with the expected utility theorem only when returns are normally distributed and the decision-maker has a quadratic utility function [279]. Mean-variance results may also be acceptable if higher moments than the second are small or relatively unimportant to the decision-maker. However, mean-variance or mean-quantile-based performance measures, such as mean-value-at-Risk, may be misleading if decision-makers want to incorporate products with nonlinear payoff profiles, such as options, in their hedging strategies. In this case, symmetrical measures of risk, like the variance or the standard deviation are not suitable, as option positions typically follow an asymmetrical risk-return-profile [52, 53]. In addition, quantile-based risk measures are not able to account for asymmetry. Furthermore, it has become a stylized fact in experimental economics that individuals perceive risk in a nonlinear fashion. Evidence suggests that most individuals perceive a low probability of a large loss to be far more risky than a high probability of a small loss [229]. This empirical finding expresses a preference towards positively skewed probability distributions.

While above studies have enhanced our understanding of what motivates individuals or corporations to manage risk, less attention has been directed towards understanding precisely how a firm should hedge. One problem is that currency risk is a rather broad and subjective concept. For example, the risk of a trader/speculator is different to a hedger's perception of risk. Furthermore, there may even exist different priorities about managing risk among hedgers. Consider the case of a multinational nonfinancial firm. First, since there naturally exist different kinds of foreign currency exposure, such as transaction exposure, translation exposure and economic exposure, all of which may be conflicting, firms may have different perceptions on what kind of exposure matters most to them. Second, hedging

---

[2] Expected utility theory states that the decision maker chooses between risky or uncertain prospects by comparing their expected utility values. This can be achieved by specifying a utility function $G : R \rightarrow R$ and then optimizing its expected value.

decisions, are often based on currency managers' fundamental beliefs as to how currency markets operate. Naturally, these beliefs will vary from individual to individual. As in all decision-making processes involving human beings, these beliefs can be engraved in cultural norms, traditions, or political factors. This means that presented with exactly the same analysis, two different people might come out with totally different interpretations of what it means to them, and what it would require in terms of action. Due to this paradigm it is practically impossible to come up with a universal answer to the question of how a firm should hedge. A large body of research, rooted in economics, international finance, portfolio theory, statistics, and operations research has explored the optimal hedging of foreign exchange risks. Due to the very specific nature of foreign exchange risk, this literature is often tailored to a very specific industry, such as agriculture, electricity, insurance, banking, or industrial firms exporting or importing goods to or from a foreign market. Depending on the nature of the context, hedging models may therefore contain very specific views, assumptions, and modeling techniques. For instance, hedging models may not only differ by sources of uncertainty and instruments permitted, but also by their formal descriptions. Since a comprehensive discussion would go beyond the scope of the dissertation, we will only summarize the kind of literature that has motivated our approach. In corporate contexts, it is often essential to use risk management to engineer cash flows. Hence, an important part of this thesis is the comparison and combination of alternative instruments and tactics for managing currency risk. We distinguish between linear contracts where the payoff of the contract is a straight line (forwards), and nonlinear contracts, where the payoff is not (options). There is no consensus in the literature regarding a universally preferable strategy to hedge currency risk, although the majority of results indicate that currency forwards generally yield better performance than single protective put options [7, 121, 282].

Important early insights into the simultaneous choice of linear and nonlinear instruments were provided by agricultural risk-management literature. Lapan et al. [242] proposed a one-period model wherein utility-maximizing managers face price risk (but not quantity risk) and choose among both forwards and options when making their hedging decisions. Assuming normally distributed prices (which allow for negative prices), they showed that the optimal hedging position will consist only of forward contracts as options become redundant. Under the assumption of nonincreasing risk-aversion and a specific strike price, they conclude that a speculative forward position is generally accompanied by a long straddle option position. Lence et al. [251] extended this model into a multiperiod framework and found that in the presence of unbiased forward and options prices, options play an important role. Another extension of the [242] model can be found in [352] who note that the optimal hedging position usually will include options, in addition to forward contracts. Reference [323] has analyzed the optimal position in forwards and straddles by comparative-static analyses plus numeric examples for various risk-utility functions. His main result is that the optimal mix of linear and nonlinear instruments depends on the specific stochastic exchange rate process that is assumed by a firm, as well as on its degree of risk aversion. Volosov et al. [408] formulate a two-stage stochastic programing decision model for foreign exchange exposure management that is based on a vector error correction model in order to predict the random behavior of the forward, as well as spot rates connecting USD and GBP. The model computes currency hedging strategies which provide rolling decisions of how much forward contracts should be bought and how much should be liquidated. Backtesting results show that the computed strategies improve the passive "spot only" strategy considerably. However their analysis does not include option instruments which are frequently used in practice. Topaloglou at al. [395] extended their own multistage dynamic stochastic programing model for international portfolio, as introduced in [394], by introducing positions in currency options to the decision set at each stage in addition to forward contracts. They find empirical evidence that portfolios with optimally selected forward contracts outperform those that involve a single protective put option per currency. However, it is also found that trading strategies involving suitable combinations of currency options have the potential to produce better performance. In summary, while many of these models are able to explain certain empirical features of hedging practices (such as partial hedging, nonlinear or dynamic strategies, etc.), none appear to explicitly account for all observed features. In particular, most models appear to come up short in two areas. First, most models do not predict substantial time-series variation in hedge ratios. Second, the models do not explicitly model the impact of managers' market views. In our opinion, these two phenomena are likely closely related and it is for this reason that we subsequently explore the potential impact of managerial views on hedging behavior.

We demonstrate how industrial corporations and investors can develop an approach to managing currency transaction risk, thereby adding real economic value from currency fluctuations. Our objective is to find a possibly optimal combination of linear and nonlinear financial instruments to hedge currency risk over a planning period such that the expected utility at the planning horizon is maximized. We require the goal function to address the conflicting empirical finding that firms do like to try to anticipate events, but that they also cannot base risk management on second-guessing the market. Our analysis therefore argues that a way to understand corporate hedging behavior is in the context of speculative motives that could arise from either overconfidence or informational asymmetries. For this purpose, we assume the firm to have different future expectations than those implied by derivative prices. If investors believe that the currency is going to move in an unfavorable direction, what should they use as a tool to hedge? If they expect the currency to move favorably, but are not entirely sure, should they use a different risk management strategy? When addressing these questions, one must recognize that traditional mean-variance or mean-quantile-based performance measures may be misleading if products with nonlinear payoff profiles, such as options, are used [52, 53]. In addition, it has become a stylized fact that individuals perceive risk in a nonlinear fashion. Preferences for positive skewness have been shown to hold theoretically by [10, 362], and empirically by, e.g., [229, 258, 372, 373]. For these reasons, we propose to embed a mean-variance-skewness utility maximization framework with linear constraints in a single-period Stochastic Combinatorial Optimization Problem (SCOP) formulation in order to find optimal combinations for forward and

European straddle option contracts given preferences among objectives. We show that the proposed SCOP is computationally hard and too difficult to be solved analytically in a reasonable amount of time. More specifically, the problem is *NP*-hard and cannot be solved in polynomial time which directly suggests that solutions can only be approximated [152, 320]. Second, computation of the objective function is nonconvex and nonsmooth. The resulting optimization problem therefore becomes fairly complex as it exhibits multiple local extrema and discontinuities.

In order to attack the given problem and derive near-optimal decisions within a reasonable amount of time, we propose a simulation/optimization procedure. Simulation/optimization has become a large area of current research in informatics, and is a general expression for solving problems where one has to search for the settings of controllable decision variables that yield the maximum or minimum expected performance of a stochastic system as presented by a simulation model [148]. For modeling the EUR/USD exchange rate a smooth transition nonlinear PPP reversion model is presented. The simulation model is very attractive in the present context and unique to our knowledge. It addresses both, the first and the second PPP puzzle, and provides a theoretically valid and visually intuitive view on the corridor of future EUR/USD spot development. The key feature is a smooth transition function [387, 403] which allows for smooth transition between exchange rate regimes, symmetric adjustment of the exchange rate for deviations above and below equilibrium, and the potential inclusion of a neutral corridor where the exchange rate does not mean revert but moves sideways. Another advantage is that it requires estimating only two parameters, namely the speed of mean reversion and exchange rate volatility.

For the task of optimization, we propose the use of a metaheuristic combinatorial search algorithm. Metaheuristics are modern heuristics which have been first introduced by [162]. For a good and recent overview of metaheuristics in combinatorial optimization, we refer to [46]. In short, a metaheuristic refers to an intelligent master strategy that guides and modifies other heuristic methods to produce solutions beyond those that are normally generated in a quest for local optimality. The specific metaheuristic we use is scatter search, a generalized form of path relinking [166, 167] which has proven to be highly successful for a variety of known problems, such as vehicle routing [18, 337, 338], tree problems [68, 423], mixed integer programing [165] or financial product design [81]. The seminal ideas of scatter search originated in the late 1960s. A first description appeared in [161], the modern version of the method is described in [239].

In order to show our simulation/optimization model's applicability in a practical context, a situation is presented, where a manufacturing company, located in the EU, sells its goods via a US-based subsidiary to the end-customer in the US. Since it is not clear what the EUR/USD spot exchange rate will be on future transaction dates, the subsidiary is exposed to foreign exchange transaction risk under the assumption that exposures are deterministic. We take the view that it is important to establish whether optimal risk management procedures offer a significant improvement over more ad hoc procedures. For the purpose of model validation, historical data backtesting was carried out and it was assessed whether the optimized mean-variance-skewness approach is able to outperform passive strategies such as unitary spot, forward, and straddle, as well as a mixed strategy over time. The passive strategies are nonpredictive since they do not allow the hedge to time-vary, but are fixed-weight strategies and therefore, do not use the history of market information. We compare the alternative strategies in dynamic backtesting simulations using market data on a rolling horizon basis. The strategies were evaluated both in terms of their ex ante objective function values – as well as in terms of ex post development of net income. We find that the optimized mean-variance-skewness strategy provides superior risk-return results in comparison to the passive strategies if earnings risk is perceived asymmetrically in terms of downside risk. The pure forward strategy is found to have the lowest return per unit of earnings risk, whereas the straddle strategy and the 1/3 strategy reveal similar risk-return characteristics.

Our results demonstrate that first, currency hedging is a hard problem from a computational complexity perspective and optimal solutions can be approximated at best. We demonstrate through extensive numerical tests the viability of a simulation/optimization model as a decision support tool for foreign exchange management. We find in our experiments that scatter search is a search method that is both aggressive and robust. It is aggressive because it finds high-quality solutions early in the search. It is robust because it continues to improve upon the best solution when allowed to search longer. Our approach to hedging foreign exchange transaction risk selectively is based on exchange rate expectations, considers real market data and incorporates flexible weights. It is found that the approach adds value in terms of reducing risk and enhancing income (which contrasts literature). Interestingly, it also contrasts the finding that currency forward contracts generally yield better results in comparison to options. The main reason for this is the forward's ability to transfer income across states of nature, concentrating the payments on the firm's downside. Thus, a smaller hedge ratio is required, which makes risk management less costly.

The chapter is organized as follows: in Chap. 12 we introduce the principles used to express preferences over probability distribution functions, including the normal distribution and the concept of expected utility maximization. In particular, Chap. 12 introduces how preferences over probability distributions can be realized using financial instruments and defines the composition of the probability distribution function to be optimized. Chapter 13 gives the exact formal representation of the currency hedging problem and describes its computational complexity. Chapter 14 describes the idea of simulation/optimization, including a formal description of the exchange rate simulation model, as well as a description of the metaheuristic optimization routine proposed. Finally, Chap. 15 describes the real-world context in which the model was applied, as well as the backtesting approach, evaluation procedure, computational tests and the empirical results.

Chapter 12
Preferences over Probability Distributions

12.1 Currency Hedging Instruments

12.1.1 Forward

A forward contract is an agreement between two parties to buy/sell an asset at a certain future time for a certain price. One of the parties to a forward contract assumes a long position and agrees to buy the underlying asset on a certain specified future date for a certain specified price. The other party assumes a short position and agrees to sell the asset on the same date for the same price. A forward contract therefore has a symmetric distribution of rights and obligations. The payoff from a long position in a forward contract on one unit of an asset is

$$F_T = \omega_T - f_{0,T} \tag{12.1}$$

where $f_{0,T}$ is the price in $t = 0$ for a forward contract with maturity T, and ω_T is the spot price of the asset at maturity of the contract. Long positions enable hedgers to protect themselves against price increases in the currency. Conversely, the payoff from a short position in a forward contract on one unit of an asset is

$$F_T = f_{0,T} - \omega_T \tag{12.2}$$

Short positions protect hedgers against price decreases. The fact that the payoffs resulting from long and short positions in forward contracts can be symmetrically positive or negative is illustrated in Fig. 12.1 [200]. The gain, when the value of the underlying asset moves in one direction, is equal to the loss, when the value of the asset moves by the same amount in the opposite direction. Since it costs nothing to enter into a forward contract, the payoff from the contract is also the firm's total gain or total loss from the contract.

A forward contract fixes the nominal exchange rate and the price until settlement. Thereby, hedging prevents transaction exposure. Since a forward contract does not hedge changes in real foreign exchange rates, economic exposure still remains.

C. Ullrich, *Forecasting and Hedging in the Foreign Exchange Markets,* Lecture Notes
in Economics and Mathematical Systems 623, DOI: 10.1007/978-3-642-00495-7_12,
© Springer-Verlag Berlin Heidelberg 2009

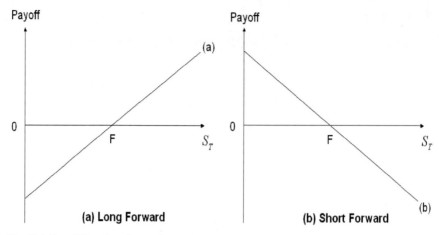

Fig. 12.1 Payoff functions from a forward contract

There can still be some foreign exchange gain or loss. When real exchange rates remain stable, long-term fixing of the nominal exchange rates can even create an economic exposure that would not have existed otherwise [332]. Since forwards can be privately arranged, that is traded over-the-counter, the terms of such a forward, for example, the maturity date and the characteristics of the underlying asset, can be customized for the users. However, liquidity and transaction costs are greater because of this customization [154].

According to the principle of risk neutral valuation, the market value of a forward contract at $t < T$ is a linear function

$$F_t = e^{-r_f(T-t)} E_t \left[\omega_T - f_{0,t} \right] \tag{12.3}$$

In $t = 0$ the forward price $f_{0,T}$ is determined such that its market value equals zero,

$$F_0 = e^{-r_f \times T} E_t \left[\omega_T - f_{0,T} \right] = 0 \tag{12.4}$$

It follows that the forward price $f_{0,T}$ equals the risk-neutral expected value of the exchange rate from a capital market view.

$$f_{0,T} = E \left[\omega_T \right] \tag{12.5}$$

Forward contracts are widely used instruments in currency hedging since they are able to inexpensively transfer risk, and reduce cashflow volatility [40, 299]. In theory, a full forward hedge is optimal, if the firm does not have any concrete expectations about the future course of the exchange rate and believes in market consensus [190, 215, 226, 322]. On the other hand, if a firm dares to deviate from risk-neutral market expectations, this would imply a speculative engagement which must result in a forward position deviating from the full hedge [422]. Typically, a selective hedger would enter into a full hedge if he expected an adverse future exchange rate development. The brighter his expectations on the future exchange rate become, the more he would reduce the hedge amount. If expectations were totally optimistic, the exposure would be left open. It is interesting that compared with theoretical results, firms in reality tend to hedge a constant amount between 0% and 100% (often 50%), if they do not have distinct expectations about the future course of the exchange rate [63, 159]. This behavior is called partial hedging and is justified by arguing that a full hedge (zero hedge) accompanied by a favorable (adverse) exchange rate development would lead to high losses, such that hedging 50% seems reasonable in the absence of concrete expectations.

12.1.2 Plain Vanilla Option

In contrast to forward contracts, options have an asymmetric distribution of rights and obligations. A call option gives the holder the right (but not the obligation) to buy the underlying asset by a certain date for a certain price. A put option gives the holder the right to sell the underlying asset by a certain date for a certain price. The price in the contract is known as the exercise price or strike price X. The date in the contract is known as the maturity. American options can be exercised at any time up to the expiration date. European options can be exercised only on the expiration date itself.

The fact that an option gives the holder the right to do something does not mean that the right must be exercised. This fact distinguishes options from forwards, where the holder is obligated to buy or sell the underlying asset. Hence, there is a fundamental difference between the use of forward contracts and options for hedging. Forward contracts are designed to neutralize transaction risk by fixing the price that the agent will pay or receive for the underlying asset. Options, in contrast, provide insurance by offering a way for agents to protect themselves against adverse price movements in the future while still allowing them to benefit from favorable price movements. The latter feature comes at an up-front fee which is referred to as the option premium P.

If we disregard the initial cost of the option, the payoff from a long position in a European call option at maturity $t = T$ is

$$C_T = \max(\omega_T - X, 0) = \begin{cases} \omega_T - X & : & \omega_T > X \\ 0 & : & \omega_T \leq X \end{cases} \tag{12.6}$$

Equation 12.6 reflects the fact that the option will be exercised if $\omega_T > X$ and will not be exercised if $\omega_T \leq X$. The asymmetric nature of a long position in a European call option is well captured by its payoff function (Fig. 12.2a). Short positions in European call options are obtained conversely (Fig. 12.2b).

The payoff to the holder of a long position in a European put option is

$$P_T = \max(X - \omega_T, 0) = \begin{cases} X - \omega_T & : & \omega_T < X \\ 0 & : & \omega_T \geq X \end{cases} \tag{12.7}$$

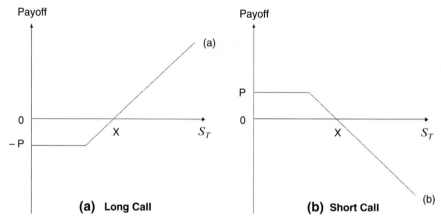

(a) Long Call   **(b)** Short Call

Fig. 12.2 Payoff functions of a European call option

If we take into account the option premium to be paid in $t = 0$, the payoff function of the call is given by the difference between C_T and the option premium multiplied by the risk free interest rate. The payoff functions for long and short positions in a European put option are depicted in Fig. 12.3a and Fig. 12.3b.

According to the principle of risk neutral valuation, the market value of European call and put option contracts at $t < T$ are given by the Black and Scholes formulae [41]

$$C_T = \omega_T \times e^{-r_f^*(T-t)}N(d_1) - X \times e^{-r_f(T-t)}N(d_2) \qquad (12.8)$$

$$P_T = X \times e^{-r_f(T-t)}N(-d_2) - \omega_t \times e^{-r_f^*(T-t)}N(-d_1) \qquad (12.9)$$

with

$$d_1 = \frac{\ln\frac{\omega_t}{X} + \left(r_f - r_f^* + \frac{\sigma^2}{2}\right)(T-t)}{\sigma\sqrt{T-t}}$$

$$d_2 = \frac{\ln\frac{\omega_t}{X} + \left(r_f - r_f^* - \frac{\sigma^2}{2}\right)(T-t)}{\sigma\sqrt{T-t}} = d_1 - \sigma\sqrt{T-t}$$

and N denoting the cumulative standard normal distribution.

Literature has recommended plain vanilla option strategies as an alternative to the selective hedging strategy in order to avoid losses that may result from having entered into a forward position when the exchange rate, ex post, took a positive direction. First, the firm has a view on future exchange rate levels and wants to participate from favorable exchange rate movements while being protected against adverse ones. In fact, it is this insurance property, i.e., the asymmetric profile, that is considered widely as the fundamental advantage of an option [40, 299]. It guarantees protection against adverse exchange rate developments while allowing participation from movements in the agent's favor. Thus, options can be motivated if a firm has

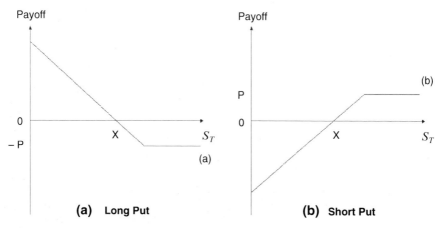

Fig. 12.3 Payoff functions of a European put option

missing expectations about the future exchange rate, if its exchange rate expectations from the past have not been successful, or if foreign exchange markets are volatile. Under homogeneous expectations about the underlying stochastic process of the exchange rate and therefore, also about its probability distribution, hedging with options cannot be optimal, since there is no deviating view from the risk-neutral value of the option. In addition, the future payoff of the option is uncertain. Hence, a full forward hedge resulting in a deterministic outcome would be optimal. Deviating expectations from those of market consensus imply a speculative position. The use of options for income enhancement has been analyzed by a streak of literature that examined whether firms incorporate exchange rate expectations in their hedging strategies [57, 237].

12.1.3 Straddle

The problem that arises when a combined position of forwards and plain vanilla option instruments has to be found is that the two instruments are coupled with complementarity and substitution between them. First, the payoff profile from a long forward can be replicated by a combined position in a long call and a short put according to put-call parity ([200], pp. 174–175). Second, both market value functions depend on the expected future exchange rate. The call option can only be distinguished from the forward by its functional dependency on the foreign exchange market's expected standard deviation of the underlying. The first-moment overlapping in the determinants of the price formation may lead to the problem that the different characteristics of forwards and options are being disguised. This problem can be avoided if straddles are used instead of plain vanilla options ([323], pp. 166–167).

A long straddle involves buying a call and a put with the same strike price and maturity date. Consequently, the payoff function of a long straddle is the result of

Fig. 12.4 Payoff function of a long straddle

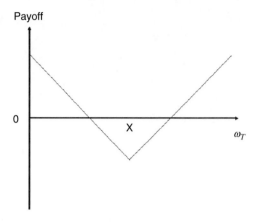

the sum of the single payoff functions of a long call and a long put.

$$S_T = |\omega_T - X| = \begin{cases} \omega_T - X & : & \omega_T > X \\ X - \omega_T & : & \omega_T \leq X \end{cases} \tag{12.10}$$

The payoff function of a long straddle is shown in Fig. 12.4. If the exchange rate is close to the strike price at maturity of the options, the straddle leads to a loss. However, for large movements in either direction, a profit can be realized. If the price goes up (down) enough, the call (put) option comes into effect and the put (call) option is ignored.

The advantage of the "buy straddle" investment tactic is that the firm still has unlimited profit potential without knowing precisely what direction the exchange rate will move in, while total loss is limited to the price of purchasing the call and put options. As long as prices do not stagnate, the firm has limited risk and a big potential for gain. Hence, the main motivation of a long straddle is to take a position on volatility play. If an upcoming news report is expected to cause an increase in volatility or rate changes for a certain currency, a straddle strategy generates better returns if the realized volatility is as expected.

12.2 Formal Relationship Between Firm and Capital Market Expectations

It is argued that a way to understand corporate hedging behavior is in the context of speculative motives that could arise from either overconfidence or informational asymmetries. For this purpose, we assume the firm to have different future expectations than those implied by derivative prices.

In the following, we assume that both the probabilistic perceptions of the firm and the risk-neutral idea of the capital market can be described by a normal distribution. Following [323], we describe deviating beliefs completely by the first two moments of the normal probability distribution. In order to model these deviations it is interesting to focus on the connection between exchange rate realizations with the same probabilities of occurrence. Formally, we refer to the connection between the exchange rate from a firm's perspective ω^u, and the risk-neutral exchange rate from a capital market's view ω^m as follows:

$$\omega^u = \omega^m + \alpha \qquad (12.11)$$

with α denoting the difference in exchange rate expectations. Differing ideas on the variability of the exchange rate under homogenuous beliefs on the expected value can be modeled by introducing a mean-preserving spread [90]:

$$\omega^u = E[\omega^u] + \varepsilon^u = E[\omega^u] + \beta \times \varepsilon^m \qquad (12.12)$$

where ε^u denotes the residual term from a firm's perspective with $E[\varepsilon^u] = 0$, ε^m denotes the residual term from capital market's perspective with $E[\varepsilon^m] = 0$, and $\beta > 0$ represents the coefficient of deviation.

From (12.11) and (12.12) we receive

$$\omega^u = E[\omega^m] + \alpha + \beta \times \varepsilon^m \qquad (12.13)$$

The uncertainty regarding the exchange rate realization from a firm's and capital market's view results from the error term ε^m, respectively, ε^u, which additively overlays the risk-neutral expected value $E[\varepsilon^m] = 0$, respectively $E[\varepsilon^u] = 0$. The deviating probabilistic perceptions between firm and capital market are captured by $\alpha \neq 0$ and $\beta \neq 1$, with $\alpha \neq 0$ representing deviating perceptions regarding the first moment, and $\beta \neq 1$ representing deviating perceptions regarding the second moment.

12.3 Specification of Probability Distribution Function

We demonstrate how industrial corporations and investors can develop an approach to managing currency transaction risk, thereby adding economic value from currency fluctuations. Empirical studies on corporate hedging behavior suggest that firms are facing a conflict. On the one hand, they like to try to anticipate events. On the other hand, they cannot base risk management on second-guessing the market. For instance, if investors believe that the currency is going to move in an unfavorable direction, what should they use as a tool to hedge? If they expect the currency to move favorably, but are not entirely sure, should they use a different risk management strategy? The following is useful for formalizing the dilemma that reduction of transaction exposure with hedging instruments directly creates economic risk.

Definition 12.1. Given a finite set S of decision variables $S = \{x_t^F, x_t^O, x_t^S\} \in [0,1]$ where x_t^F, x_t^O, and x_t^S denote the forward, straddle option, and spot weights with $x_t^F + x_t^O + x_t^S = 1$, and $t = 1, \ldots, T$. Let the discrete probability distribution function of transaction exposure be

$$v(x, w) = \sum_{t=1}^{T} z_t \times (x_t^S \times \omega_t + x_t^F \times f_{0,t} + x_t^O \times X_{0,t}) \tag{12.14}$$

and the discrete probability distribution function of economic exposure be

$$w(x, \omega) = \sum_{t=1}^{T} z_t (x_t^F \times (\omega_t - f_{0,t}) \tag{12.15}$$
$$+ x_t^O \times (|\omega_t - X_{0,t}| - p_{0,t} \omega_{t=0} - c \times (x_t^F + x_t^O))$$

where z denotes the transaction exposure and c the transaction cost for a single transaction as imposed by the counterparty. The expressions in brackets, as given in (12.15), denote the payoffs of long forward and straddle option contracts which are used to reduce transaction exposure in (12.14). The random variable at t is given by the spot exchange rate ω_t, $t = 1, \ldots, T$. For purposes of this analysis, it is assumed that an investor holds one unit of foreign exchange, i.e., $z_t = 1$. The effects of marking-to-market and margin requirements are disregarded since they are purely accounting related and do not impact economic performance.

12.4 Expected Utility Maximization and Three-Moments Ranking

12.4.1 Preference Structures over Lotteries

The objective of finding a possibly optimal combination of linear and nonlinear financial instruments to hedge currency risk over a planning period requires formulating a goal function incorporating (12.14) and (12.15), as well as defining a measure of preference in order to evaluate the quality of a feasible solution. In decision theory, utility is a measure of the desirability of consequences of courses of action in a decision made under uncertain conditions. Utility theory was formalized mathematically by the classic work of [411]. Von Neumann and Morgenstern introduced a set of necessary and sufficient axioms for the rational decision-maker, and promoted the development of methods to measure utilities on numerical scales, resulting in real numbers representing personal values – such that one alternative with probabilistic consequences is preferred to another if and only if its expected utility is greater than that of the alternative. The axiomatic system is based on preference structures and their numerical representations in form of utility functions which will be described in the following.

Definition 12.2. A preference structure can be formulated as a binary preference relation. Suppose Z is a set of outcomes resulting from an agent's decisions. We assume that the agent's preference structure over Z is captured by a preference relation $\succ$ on Z, defined as a binary relation with the following property: for all outcomes $A, B \in Z$

- A is preferred over B if and only if $A \succ B$
- A is indifferent to B if and only if $A \sim B$, that is neither $A \succ B$ nor $A \prec B$
- $A \succeq B$ if and only if $A \succ B$ or $A \sim B$

Lemma 12.1. *The preference relation $\succeq$ is a total order, that is, every pair of outcomes are comparable. Formally, the relation $\succeq$, as induced by $\succ$ and thus $\sim$, is a binary relation with the following three properties:*

- *Reflexivity: For all outcomes $A \in Z$, $A \sim A$*
- *Orderability: For all outcomes $A, B \in Z$, exactly one of the following holds $A \succ B$, $B \succ A$, or $A \sim B$*
- *Transitivity: For all outcomes $A, B, C \in Z$, if $A \succeq B$, and $A \succeq C$, then $B \succeq C$.*

Preference structures can be extended to uncertain outcomes. Uncertain outcomes are represented using (discrete) probability distributions over Z, referred to as lotteries.

Definition 12.3. A lottery L with a finite number of distinct outcomes $C_1, C_2, \ldots, C_n \in Z$ and respective probabilities $p_1, p_2, \ldots, p_n$ is written as $L = \{(p_1, C_1), (p_2, C_2), \ldots, (p_n, C_n)\}$ where $\sum_{i=1}^{n} p_i = 1$, $p_i \geq 0$, $i = 1, \ldots, n$.

Corollary 12.1. *A lottery is degenerate if $n = 1$, in which case the probability in the notation can be omitted. A lottery can also have a countably infinite number of distinct outcomes $C_1, C_2, \ldots, C_n, \ldots \in Z$ with respective probabilities $p_1, p_2, \ldots, p_n, \ldots$ which can be written $L = \{(p_1, C_1), (p_2, C_2), \ldots, (p_n, C_n), \ldots\}$ where $\sum_{i=1}^{\infty} p_i = 1$, $p_i \geq 0$, $i = 1, \ldots, n, \ldots$.*

We denote as $L(Z)$ the set of all lotteries over the outcome set Z. The agent's preference relation is correspondingly extended to lotteries, and we still use the notation $\succ$ for the preference relation on lotteries.

Lemma 12.2. *Preference structures over lotteries have additional properties that are consistent with human intuition about preference structures involving uncertainty:*

- *Continuity: For all lotteries $A, B, C \in Z$, if $A \succ B \succ C$, then there exists $p \in [0, 1]$ such that $((p, A), (1 - p, C)) \sim B$*
- *Substitutability: For all lotteries $A, B \in L(Z)$, if $A \sim B$ then for all lotteries $C \in L(Z)$ and all $p \in [0, 1]$, $((p, A), (1 - p, C)) \sim ((p, B), (1 - p, C))$*
- *Monotonicity: For all lotteries $A, B \in L(Z)$, if $A \succ B$ then for all lotteries $p, q \in [0, 1]$, $p \geq q$ if and only if $((p, A), (1 - p, B)) \sim ((q, B), (1 - q, B))$*
- *Decomposability: For all lotteries $A, B, C \in Z$ and all $p, q \in [0, 1]$, $((p, A), (1 - p, [(q, B), (1 - q, C)])) \sim ((p, A), ((1 - p)q, B), (1 - p)(1 - q), C)).$*

These properties imply that the extended preference relation is also a total order on $L(Z)$. We will refer to the above set of seven properties as the utility axioms.

12.4.2 Preference Structures over Utility Functions

For practical decision problems under uncertainty, however, it is more convenient to define preference structures based on real-valued functions on $L(Z)$.

Lemma 12.3. *A real-valued function on $L(Z)$ can define a preference structure satisfying the utility axioms if and only if it satisfies the functional form of the utility axioms, namely*

- Continuity: For all lotteries $A, B, C \in L(Z)$, if $U(A) > U(B) > U(C)$, then there exists $p \in [0,1]$ such that $U((p,A),(1-p,C)) = B$
- Substitutability: For all lotteries $A, B \in L(Z)$, if $U(A) = U(B)$ then for all lotteries $C \in L(Z)$ and all $p \in [0,1]$, $U((p,A),(1-p,C)) = U((p,B),(1-p,C))$
- Monotonicity: For all lotteries $A, B \in L(Z)$, if $U(A) = U(B)$ then for all lotteries $p,q \in [0,1]$, $p \geq q$ if and only if $U((p,A),(1-p,B)) \sim U((q,B),(1-q,B))$
- Decomposability: For all lotteries $A, B, C \in L(Z)$ and all $p,q \in [0,1]$, $U((p,A),(1-p,[(q,B),(1-q,C)])) = U((p,A),((1-p)q,B),((1-p)(1-q),C))$

Since the relation $\geq$ is a total order on R, the first three utility axioms: reflexivity, orderability, and transitivity do not need to be included.

12.4.3 Expected Utility Maximization

The underlying assumption in utility theory is that the decision maker always chooses the alternative for which the expected utility is maximized. To determine the expected utility, a utility value has to be assigned to each of the possible consequences of each alternative. A utility function maps utility to the range of outcomes of a decision, depending on the decision maker's preferences and attitude toward risk. Utility theory is therefore intrinsically related to the concepts of risk and uncertainty in decision making.

Lemma 12.4. *If the extended preference relation satisfies the utility axioms, then there exists a utility function $U : L(Z) \to R$ such that for all lotteries $A, B \in L(Z)$, $A \succeq B$ if and only if $U(A) \geq U(B)$ where the utility of a lottery L is the expectation of the utilities of its components*

$$U(L) = U((p_1,C_1),(p_2,C_2),\dots,(p_n,C_n),\dots) = \sum_{i=1}^{\infty} p_i \times U(C_i) = E[U(z)] \quad (12.16)$$

where z is the random variable for the outcomes of lottery L.

Therefore, the utility function only needs to be defined for elements in Z, and its definition on $L(z)$ can be obtained using (12.16).

Definition 12.4. A is preferred to B if and only if terminal wealth satisfies $E[U(z_A)] - E[U(z_B)] \geq 0$ with at least one strict inequality $U(z_A) - U(z_B) \geq 0$.

It must be emphasized, by convention, utility is purely an ordinal measure. In other words, utility can be used to establish the rank ordering of outcomes, but cannot be used to determine the degree to which one is preferred over the other. For example, consider two outcomes A and B with corresponding utilities of 100 and 25. We can say that A is preferred over B, but we cannot say that A is four times more preferred than B. As a consequence of this ordinality, utility functions unique up to a positive linear transformation, that is, both $U_1(\cdot)$ and $U_2(\cdot) = a \times U_1(\cdot) + b$ (where $a > 0$), are utility functions corresponding to the same preference structure.[1]

12.4.4 Increasing Wealth Preference

Firms can have different risk attitudes, that is, different ways of evaluating a (non-degenerate) lottery. Risk attitudes can be modeled using utility functions. Based on the utility function, different types of risk attitudes can be identified [139].

Definition 12.5. A utility function possesses increasing wealth preference if and only if $U'(z) \geq 0$ for all z with at least one strict inequality.

This feature captures the "more wealth is better" (non-satiability) philosophy of firm behavior and is generally considered a universal feature of utility functions. For greater wealth to be preferred, the utility function must be monotonically increasing.

12.4.5 Risk Aversion Preference

This feature captures the willingness of a firm to purchase insurance (i.e., to pay more than the expected loss to transfer an insurable loss). This is a subset of increasing wealth preference, a firm may have increasing wealth preference with or without exhibiting risk aversion, and is also generally considered a universal feature of utility functions. Mathematically this is expressed as:

Definition 12.6. A utility function possesses risk aversion if and only if it satisfies the conditions for increasing wealth preference and $U''(z) \leq 0$ for all z with at least one strict inequality.

This mathematical definition of risk aversion is equivalent to the behavioral definition given above. Note, that Definition 12.6 defines a concave function. By applying Jensen's inequality this yields $E[U(z)] \leq U(E[z])$, i.e., under risk aversion the expected utility of a risky investment is less than the utility of the expected outcome. The reason for this phenomenon is that, by proposition, the firm has penalized the utility of the investment for the possibility of unfavorable outcomes. If we rewrite Jensen's inequality with a strict inequality we can show that $E[U(z)] = U(E[z] - \pi)$.

---

[1] For proof of this proposition, see [179].

This shows that the firm is indifferent between the return on a risky investment or a lower, risk-free wealth equal to $E[z] - \pi$ where π is the premium that the firm is willing to pay to eliminate risk.

Corollary 12.2. *If $E[U(z)] = U(E[z])$ for all z, then the utility function is risk-neutral. Otherwise, if $E[U(z)] > U(E[z])$ for all z, then the utility function is risk-seeking.*

Pratt [330] defined local properties of risk attitudes, as well as a measure of risk sensitivity, which are convenient for risk attitudes involving both risk aversion and risk seekingness, such as the risk attitude of an agent who buys insurance and lottery tickets at the same time. The local risk property is related to the local convexity of the utility function.

Definition 12.7. If the utility function is twice differentiable and strictly monotonically increasing then the degree of absolute risk aversion is given by a local risk measure

$$\lambda(z) = -\frac{U''(z)}{U'(z)} \tag{12.17}$$

which relates to the global risk properties as follows: if $\lambda(z) = 0$ everywhere, then the utility function is linear, and thus the agent is risk-neutral. If $\lambda(z) > 0$ everywhere, then the utility function is strictly concave, and thus the agent is risk-averse. If $\lambda(z) < 0$ everywhere, then the utility function is strictly convex, and thus the agent is risk-seeking.

12.4.6 Ruin Aversion Preference

Investors typically distinguish between upside and downside risks implying that more than the mean and variance of returns is priced in equilibrium. Negatively skewed probability distributions imply that the risk for a substantial loss is bigger than the chance of a substantial gain, and firms are hence, averse to it. Positively skewed distributions imply that the chance of a substantial gain is bigger than the risk for a substantial loss which is typically a desirable state for an investor or a firm. Preferences for positive skewness have been shown to hold theoretically by [10, 362], and empirically by, e.g., [258,372,373]. The preference for positive skewness is classically presented as an individual's willingness to play the lottery: to accept a small, almost certain loss in exchange for the remote possibility of huge returns. A firm's concern, however, is with the opposite situation: unwillingness to accept small, almost certain gain in exchange for the remote possibility of ruin [229]. The logic is therefore that a firm would be ready to trade off some average return for a lower risk of high negative returns. Ruin aversion is a subset of risk aversion. A firm may have risk aversion with or without exhibiting ruin aversion.

Definition 12.8. A utility function possesses ruin aversion if and only if it satisfies the conditions for risk aversion and $U'''(z) \geq 0$ for all z with at least one strict inequality.

As with risk aversion, it is not intuitively clear that the mathematical and behavioral definitions of ruin aversion are consistent. Formally, a positive skewness preference is related to the positivity of the third derivative of the utility function. This can be illustrated by taking a Taylor series expansion of the expected utility

$$
\begin{aligned}
U(z) = U(E[z]) + U'(E[z]) \times (z - E[z]) \\
+ \frac{U''(E[z])}{2!} \times (z - E[z])^2 \\
+ \frac{U'''(E[z])}{3!} \times (z - E[z])^3
\end{aligned}
\tag{12.18}
$$

which leads to the expression

$$
E[U(z)] = U(E[z]) + \frac{U''(E[z])}{2!} \times Var[z] + \frac{U'''(E[z])}{3!} \times Sk[z]
\tag{12.19}
$$

From this expression it can be seen that $U''(E[z])$ and $U'''(E[z])$ are, respectively, related to variance and skewness. While a negative second derivative $U'''(E[z]) < 0$ of the utility function implies variance aversion, a positive third derivative of the utility function $U'''(E[z]) > 0$, entails a preference for positive skewness. Hence, any investment feature that decreases variance and increases positive skewness (or reduces negative skewness) acts to increase expected utility.

Proof. From Definition 12.7, decreasing absolute risk aversion or constant absolute risk aversion as an upper bound is present, if

$$
\frac{\partial \lambda(z)}{z} = \frac{-u' \times u''' + (u'')^2}{(u')^2} \leq 0
\tag{12.20}
$$

which implies $u''' > (u'')^2 \times (u')^{-1} > 0$ in case of decreasing absolute risk aversion and $u''' = (u'')^2 \times (u')^{-1} > 0$ in case of constant absolute risk aversion.   □

Nondecreasing absolute risk aversion therefore implies a preference for positive skewness (lower negative skewness) on the distribution of wealth.

12.5 Specification of Utility Function

Like many other real world problems, our problem requires the simultaneous optimization of multiple competing criteria. Utility theory proposes to compute solutions to such problems by combining them into a single criterion to be optimized, according to some utility function. For half a century, Markowitz's two-moment mean-variance model has been the default model in financial engineering and the benchmark model for new theories of portfolio choice [278–281]. In this model, the choice of asset allocations is asserted to depend solely on the expected return (mean) and risk (variances and covariances) of the admissible assets. From the outset, portfolio selection was conceived as a two-step procedure: the determination of the efficient set of portfolios as the first step, preparing the selection of an optimal port-folio for a given preference structure in the second step. The idea of utility maxi-mization as a methodology for portfolio optimization problems, based on the utility theory founded by [411], can be traced back at least to [391], and also appeared in the early assessments of the mean-variance approach (e.g., [257, 347, 348]). Risk-sensitive utility functions are defined in order to identify a (possibly) optimal portfolio. Levy and Markowitz [257] showed for various utility functions and empir-ical returns distributions, that the expected utility maximizing agent could typically do very well if he acted knowing only the mean and the variance of a probability distribution. This makes the model simple to apply, but it is based on the assumption that either (1) the return distribution is normal or that (2) investors are indifferent to higher moments and equally averse to downside and upside risk.[2] Regarding the first argument there has been extensive debate: in reality, financial returns rarely are normal – more often than not, skewness and kurtosis deviate from normality, mak-ing the mean-variance approximation unlikely to select the true optimal portfolio [250, 273]. With regards to the second argument, it has become a stylized fact that individuals perceive risk in a nonlinear fashion. Apart from a firm's natural desire to avoid large economic losses from derivative hedging, we motivate the skewness criterion for a more practical reason: empirical literature on the hedging behavior of firms states that both currency forward and option vehicles are used to hedge against currency risk. However, probability distributions cannot be formally distin-guished by their first and second moments, if nonlinear instruments such as options [52, 53] and/or dynamic strategies are considered [189, 350]. As a result, we extend the concept of mean-variance trade-off to include the skewness criterion in the cur-rency hedging problem. A kurtosis criterion is not incorporated due to the above mentioned fact that a consideration of long horizon returns does usually not involve significant kurtosis.

Definition 12.9. Let the standard measure of return be defined by mean, and the standard measure of risk be defined by variance and skewness. With respect to Definition 12.8, this results in a three-order polynomial utility function

$$U(v,w) = E[w] - \{\lambda Var[v] - \kappa Sk[w]\} \tag{12.21}$$
$$= E[w] - \{\lambda E[(v - E[v])^2] - \kappa E[([w] - E[w])^3]\}$$

where $\lambda > 0$ denotes the coefficient of absolute risk aversion, and $\kappa > 0$ denotes the coefficient of absolute prudence.

Equation 12.21 is a multiobjective cubic function defined over the random variable ω. As [200] notes on pp. 240–241, the expected rate of return is an ambiguous concept. Unless otherwise stated, it will be used to refer to $E[w] = \mu$. The maximum

---

[2] These limitations of the mean-variance approach were established at an early point by, e.g., [12,75,129,182,183,279,330,391]. Rothschild and Stiglitz [347] showed from several perspectives that the usage of variance as a definition of risk is insufficient. More recent assessments of the mean-variance approach include, e.g., [76,77,272,293].

value function for the firm is then determined for a given set of parameters $(\lambda, \kappa) > 0$ representing the degree for absolute risk aversion and absolute prudence. Typical values of absolute risk aversion and prudence have been suggested to lie in the range of 1–5 [110, 111, 290].

Equation 12.21 is independent of a firm's wealth level. In particular, a change in wealth just causes a parallel shift for the model which will not affect the risk attitude and the choice behavior of the firm. For symmetric distributions or distributions not highly skewed, such that $E[(v - E[v])^2] > cE[(w - E[w])^3]$, the model will be risk-averse. However, for highly positively skewed distributions, such that the skewness of the distribution of derivative payoffs overwhelms the variance of the distribution of transaction exposure , i.e., $E[(v - E[v])^2] < cE[(w - E[w])^3]$, the model will be ruin-averse.

Chapter 13
Problem Statement and Computational Complexity

13.1 Problem Statement

In order to find possibly optimal combinations between spot, forward, and European straddle option contracts, it is proposed to embed the three-moment utility function as formulated in Definition 12.9 in a single-period stochastic combinatorial optimization problem (SCOP) framework with linear constraints. Decisions are made solely at $t = 0$ and cannot be revised in subsequent periods. The decision horizon is finite and set to $T = 1$. Due to the current popularity of multiperiod financial models, a single-period description of the problem might be considered as disputable. In a dynamic modeling context, decisions are optimized in stages, because uncertain information is not revealed all at once. In our context this would require building an integrated dynamic simulation model that additionally describes the future development of the forward rates as well as the option premiums over the next 12 months in a stochastic manner. Obviously, such an approach would be much more difficult to model than our single-factor approach, contain significantly higher model risk (three risk factors plus interactions), and would require more computational resources. It is therefore doubtful if such a model would indeed be advantageous, especially if we consider the underlying decision context. In practice, firms' currency hedging decisions are hardly enforced by speculative behavior towards future price developments of hedging vehicles. Instead, they rather concentrate on reward/risk estimates that depend solely on the underlying risk factor, i.e., the exchange rate in our case. Any further assumptions or risks are usually avoided. Hence, although a multiperiod decomposition of reward and risk poses an interesting theoretical problem, we argue that it is less relevant in practice. Therefore, only forward rates and option prices as known in $t = 0$ are relevant. Future prices of hedging vehicles are irrelevant and thus, not considered. The single-period SCOP is stated as follows:[1]

Problem 13.1. (BEST HEDGE). Given a probability space (Ω, Σ, P), where Ω is the domain of random variables ω (typically a subset of R^k), Σ is a family of *events*,

---

[1] For the general definition see [216].

C. Ullrich, *Forecasting and Hedging in the Foreign Exchange Markets,* Lecture Notes in Economics and Mathematical Systems 623, DOI: 10.1007/978-3-642-00495-7_13, © Springer-Verlag Berlin Heidelberg 2009

that is subsets of Ω, and P is a probability distribution Σ with $P(\Omega) = 1$. Consider
also a finite set S of decision variables x. S is typically a subset of R^n. Given a finite
set of feasible solutions x, a real-valued utility function $U(v(x,\omega),w(x,\omega))$ of the
two variables $(x,\omega) \in (S,\Omega)$ and constraint functions H_i, $i = 1,2,\ldots,m$ mapping
$(x,\omega) \in (S,\Omega)$ to R find

$$\max_{x \in s} \quad U(v(x,\omega),w(x,\omega)) \tag{13.1}$$

$$\text{subject to} \quad H_i(w,\omega) \leq 0, \quad i = 1,\ldots,m$$

where, $U(v(x,\omega),w(x,\omega))$ is given by

$$U(v,w) = E[w] - \{\lambda Var[v] - \kappa Sk[w]\} \tag{13.2}$$

$$= E[w] - \{\lambda E[(v - E[v])^2] - \kappa E[(w - E[w])^3]\}$$

according to Definition 12.9, and

$$v(x,w) = \sum_{t=1}^{T} z_t \times (x_t^s \times \omega_t + x_t^F \times f_{0,t} + x_t^O \times X_{0,t})$$

$$w(x,\omega) = \sum_{t=1}^{T} z_t(x_t^F \times (\omega_t - f_{0,t}) \tag{13.3}$$

$$+ x_t^O \times (|\omega_t - X_{0,t}| - p_{0,t}\omega_{t=0}) - c \times (x_t^F + x_t^O))$$

according to Definition 12.1.

Single-period SCOP formulations have been originally proposed in the context of
mathematical programing applied to SCOPs, and this field is also called in the literature stochastic integer programing (SIP), a subset of the broader field of stochastic
programing [38]. Surveys on SIP include [181,225]. Single-period SCOPs are characterized by the fact that decisions, or equivalently, the identification of a possibly
optimal solution is done before the actual realization of the random variables. This
framework is applicable when a given solution may be applied with no modifica-
tions (or very small ones) once the actual realization of the random variables are
known.

Global optimization is the task of finding the absolutely best set of decision variables to optimize its objective function. In general, there can be solutions that are
locally optimal but not globally optimal.

Definition 13.1. A local maximizer x^* of $U(v(x,\omega),w(x,\omega))$ is a point such that
there exists a neighborhood B of x^* with:

$$U(v(x^*,\omega),w(x^*,\omega)) \geq U(v(x,\omega),w(x,\omega)), \forall x \in B \tag{13.4}$$

Many local maxima may exist with substantially different function values. For problems with multiple maxima we may be interested in finding the best maximum.

Definition 13.2. The global maximization problem for a function $U(v(x^*,\omega),$ $w(x^*,\omega)) \geq U(v(x,\omega),w(x,\omega))$ is to find x^* such that:

$$U(v(x^*,\omega),w(x^*,\omega)) \geq U(v(x,\omega),w(x,\omega)), \; \forall x \in S \qquad (13.5)$$

All practical optimization methods, whether local or global, are iterative in nature and typically proceed from a starting point, which is an estimate of x^*, via a sequence of points with increasing function value, until some termination condition is satisfied. Local optimization problems can be solved more easily than global ones since a local solution can be characterized by computable information at x^* (positive definiteness of the Hessian and zero gradient), whilst for the case of a global optimum no such criteria exist, in general. The aim of global optimization is to find the points in S for which the function reaches its highest value, the global maximum.

In order to investigate the computational complexity of Problem 13.1, we focus on the theory of *NP*-completeness [152, 320]. This theory is designed to be applied only to decision problems, i.e., problems whose solution is either "yes" or "no." The standard formulation of a decision problem consists of two parts: the first part specifies a generic instance of the problem in terms of various components such as sets, graphs, functions, numbers, etc. The second part states a yes/no question asked in terms of the generic instance.

Problem 13.2. (BEST HEDGE Decision Problem). Given a probability space (Ω, Σ, P) a finite set of feasible solutions x, a real-valued utility function $U(w(x,\omega))$ of the two variables $(x,\omega) \in (S,\Omega)$, and an integer bound B, is there a solution x such that

$$u(x) := E_P(U(v(x,\omega),w(x,\omega))) \leq B? \qquad (13.6)$$

(That is, is there a vector x such that a given utility target level can be realized?)

13.2 Computational Complexity Considerations

13.2.1 Complexity of Deterministic Combinatorial Optimization

If the exchange rate over the planning horizon was known today, then Problem 13.1 would correspond to a deterministic combinatorial optimization problem (DCOP). All information would be available at the decision stage and could be used by an optimization algorithm to find a possibly optimal solution. The concrete application of a solution found would lead exactly to the cost of the solution as computed by the optimization algorithm. Therefore DCOPs are also considered static problems because from the point of view of the decision maker, there is nothing else to be decided after the optimization took place.

Note that in Problem 13.1 the size of the choice set has already been reduced by permitting only two kinds of hedging instruments. The size of the true choice set in Problem 13.1 is determined by the number of permitted hedging instruments n, the degree of discretization d, the speculation constraint, and the length of the planning horizon T. In absence of speculation and possible budget constraints, a collection of $m = T \times (n \times (d+1))!$ binary decisions is given, each regarding the purchase of a combination of instruments at price p. With budget I the firm faces a subset of I/p combinations out of the complete set of binary decisions. The more budget is at the firm's disposal, the more "optionality" can be added to standard forward hedges, and thus the more combinations must be considered. However, absence of a budget constraint does not mean that optionality can be added infinitely. It is the speculation constraint that restricts the firm in purchasing more forwards/options than are actually required to close the open positions. The speculation constraint therefore represents a bound on the number of combinations to be considered. At the maximum, the firm must answer m binary decision problems which takes time proportional to $(n \times d)!$ which is not a polynomial bound.

Proposition 13.1. *The deterministic version of Problem 13.1 is NP-hard as the MINIMUM COVER Problem is a special case of the deterministic decision version of Problem 13.1.*

Proof. The result can be proven in a similar way as [156] prove their Consumer Problem to be *NP*-complete. This is done by reducing the problem MINIMUM COVER ([SP5]) as represented in [152], p. 222, to the deterministic version of Decision Problem 13.1.   □

Problem 13.3. (MINIMUM COVER). Given a collection C of subsets of a finite set S, and a positive integer $K \leq |C|$, does C contain a cover for S of size K or less, that is, a subset $C' \subseteq C$ with $C' \leq |K|$ and such that $\bigcup_{c \in C'} c = S$?

Let there be given an instance of MINIMUM COVER: A finite set $S \equiv \{1, \ldots, r\}$ of length $|S| = r$, a collection of subsets $C = \{S_1, \ldots, S_q\}$ of length $|C| = q$, and a positive integer $k \leq q$. Let $(y_{ij})_{i \leq q, j \leq r}$ denote the incidence matrix, namely $y_{ij} = 1$ if $j \in S_i$ and $y_{ij} = 0$ if $j \notin S_i$.

We now define the associated hedging problem. Let $n = q$. For $i \leq n$, let $p_i = 1$, and define $I = t$. Next, define $u(x)$ by $u(x_1, \ldots, x_n) = \Pi_{j \leq r} \sum_{i \leq n} y_{ij} x_i$. Finally, set $\bar{u} = 1$. A bundle satisfies $\sum_{i \leq n} p_i x_i \leq 1$ and $u(x_1, \ldots, x_n) \geq \bar{u}$ if and only if $\sum_{i \leq n} x_i \leq t$ and $\sum_{i \leq n} y_{ij} x_i \geq 1$ for every $j \leq r$. In other words, the firm has a feasible bundle $x \equiv (x_1, \ldots, x_n)$ obtaining the utility of 1 if and only if:

1. No more than t products of $1, \ldots, n$ are purchased at a positive quantity at x, and
2. The subsets S_i corresponding to the positive x_i form a cover of $S = \{1, \ldots, r\}$.

The construction above can be performed in linear time. It is then left to show that we have obtained a legitimate utility function u. Continuity holds because this is a well-defined function that is described by an algebraic formula. Since $y_{ij} \geq 0$, u is nondecreasing in the x_i's. We turn to prove that it is quasi-concave. If there exists $j \leq r$ such that $y_{ij} = 0$ for all $i \leq n$, $u(x_1, \ldots, x_n) = 0$, and u is quasi-concave. Let us therefore assume that this is not the case. Hence u is the product of r expressions, each of which is a simple summation of a nonempty subset of $\{x_1, \ldots, x_n\}$. On the domain $\{x | u(x) > 0\}$, define $v = \log(u)$. It is sufficient to show that $v(x_1,\ldots,x_n) = \sum_{j \leq r} \log(\sum_{i \leq n} y_{ij} x_i)$ is quasi-concave. Obviously, for every $j \leq r$, $\log(\sum_{i \leq n} y_{ij} x_i)$ is a concave function, and the sum of concave functions is concave. This completes the proof of Proposition 13.1.

Thus (assuming $P \neq NP$), it is impossible for a nondeterministic Turing machine to verify in polynomial time whether there exists a bundle of instruments such that a firm can obtain a given level of utility without violating its speculation constraint. The result suggests that it is hard to maximize a utility function over a large choice set – even in the absence of uncertainty. As rational as investors can possibly be, it is unlikely that they can solve, in their minds, problems that prove intractable for computer scientists equipped with the latest technology. Correspondingly, it is always possible that a firm will fail to even consider some of the bundles available to it. In addition, utility maximization may not give an answer on how firms choose a bundle in the feasible set. It follows, that one cannot simply teach firms to maximize their utility functions.

13.2.2 Complexity of Stochastic Combinatorial Optimization

Realistic utility maximization problems, such as Problem 13.1, are likely to be burdened with two sources of difficulty: first, the utility function may not be known for many bundles that have not been consumed. In this context, psychological literature suggests that people do not seem to be particularly successful in predicting their own well-being as a result of future consumption. That is, consumers do not excel in *affective forecasting* (see [155, 214]). Firms may be uncertain about their utility functions, and they may learn them through the experience of hedging. In this sense, a firm is faced with a familiar trade off between exploration and exploitation: trying new options in order to gain information and selecting among known options in an attempt to use this information for maximization of well-being. By contrast, Problem 13.1 ignores this difficulty of learning the utility function. We assume that the utility function is given, as an easily applicable formula, and that, given a particular bundle, there is uncertainty regarding the utility derived from it due to the existence of uncertainty about the future exchange rate.

One possibility with problem formulations involving uncertainty is to describe uncertain information by means of random variables of known probability distributions. Under this assumption, the optimization problem is stochastic, and the objective function strongly depends on the probabilistic structure of the model. The advantage of using SCOPs over DCOPs is that the solutions produced may be more easily and better adapted to practical situations where uncertainty cannot be neglected. However, this comes at a price: first, for a practical application of SCOPs, there is the need to assess probability distributions from real data or subjectively, which is a task that is far from trivial. Second, the objective function is typically much more computationally demanding in SCOPs than in DCOPs. Although the problem is formulated as a single-period SCOP, one can distinguish a time before the actual realization of the random variables, and a time after the random variables are revealed, because the associated random events happen. Hence, our problem formulation is consistent with the probabilistic combinatorial optimization problem (PCOP) framework as presented in [34, 35]. In PCOPs, one is confronted with a valid probabilistic data set in the first stage. In a second stage, the actual data set materializes. There are two possible strategies for solving this sort of problem: reoptimization and a priori optimization which have both been shown to be *NP*-complete problems [34, 35].

13.2.2.1 Computational Complexity of Reoptimization

In the above application, one finds that after solving a given instance of a combinatorial optimization problem, it becomes necessary to solve repeatedly many variations of the instance solved originally. The most obvious approach in dealing with such cases is to attempt to solve optimally (or near-optimally) every potential instance of the original problem. Such an approach is called a reoptimization strategy. Similar to [34], p. 91, who considered the probabilistic traveling salesman problem (PTSP), it can be shown that the proof of this problem is straightforward.

Problem 13.4. (PTSP1). Given a set of instances D from a complete graph $G = (V, E)$, $|V| = n$, a cost $f : E \rightarrow R$, probability $p(d)$ that the single instances materialize and a bound B, does there exist a tour τ such that

$$E[L_\tau] = \sum_{d \subseteq D} p(d) L_\tau(d) \leq B? \tag{13.7}$$

If every possible subset of the node set V may or may not be present on any given instance of the optimization problem then there are 2^n possible instances of the problem, namely all the possible subsets of V. Hence, one has to compute $O(2^n)$ terms, each of which requires the evaluation of a classical TSP problem which is known to be *NP*-complete (see Problem [ND22] in [152], pp. 211–212). Thus, even in the simple case where we assume binary scenarios, an exponential number of classical TSPs has to be solved. Our underlying problem represents a generalization of this idea including stochastic demands which are not only binary (demand of one unit with a certain probability), but can be any random variable. Nevertheless, the problem size will be exponential in n.

13.2.2.2 Computational Complexity of A Priori Optimization

The reoptimization approach suffers from two disadvantages. Since the combinatorial optimization problem considered is *NP*-complete, one has to solve exponentially many instances of a hard problem. However, in financial applications it is usually necessary to find a solution to each new instance quickly, but one might not have the required computing power or other resources for doing so. A priori optimization is a strategy that differs from reoptimization (see [35, 204, 243]). Given a set of first-stage probabilistically valid input data, one optimizes a priori over this data in order to be able to efficiently modify this solution as needed to fit the actually realized second-stage data. Let D denote the first-stage probabilistically valid data set and $X(D)$ denote the a priori solution on D. Furthermore, let M denote a modification strategy which modifies the a priori solution so that it fits the second-stage realized data. The value of the solution output by M for an a priori solution $X(D)$ and a materialized valid data set $d \subseteq D$ is then $M(X(D),d)$. The a priori optimization rule requires that we obtain an a priori solution $X(D)$ over D such that the expected outcome of the modification strategy is minimized

$$E[M] = \sum_{d \subseteq D} p(d)M(S(D),d) \qquad (13.8)$$

It is important to note three points in this context. First, PCOP problems, such as the PTSP, differ with respect to the particular modification strategy M. Second, although the design process of solving a PCOP problem is a two-stage procedure, the robustness criterion of an optimum expectation pertains only to the second stage outcome. Third, optimizing $E[M]$ essentially requires probabilistic analysis of the modification strategy M over a distribution of input data. The decision problem of the a priori optimization version of the PTSP can be expressed as follows.

Problem 13.5. Given a set of instances D from a complete graph $G = (V,E)$, $|V| = n$, a cost $f : E \rightarrow R$, probability $p(d)$ that the single instances materialize, an updating method M, and a bound B, does there exist a tour τ of length ("cost") L such that

$$E[L_{M(\tau)}] = \sum_{d \subseteq D} p(d)L_{M(\tau)}(d) \leq B? \qquad (13.9)$$

In other words, the goal is to minimize the "weighted average" over all problem instances of the values $L_{M(\tau)}(d)$ obtained by applying the updating method M to the a priori solution t.

Theorem 13.1. [34]. *Problem PTSP2 is NP-complete.*

Noticeably, no general approximation technique has emerged under the PCOP framework. This is mainly due to the fact that PCOP concerns probabilistic analysis of a modification strategy over a distribution of input data, given an a priori solution. Therefore, analysis is always heavily dependent on the characteristics of the modification strategy used, as much as on the structure of the underlying deterministic problem studied under the PCOP framework.

13.2.3 Objective Function Characteristics

Apart from structural reasons concerning the problem as a whole, the decision to incorporate skewness in evaluating hedging instruments makes the optimal selection procedure more complicated than a mean-variance based evaluation procedure.

Whereas, the latter one implies evaluating a quadratic function, the mean-variance-skewness objective function is cubic, and practical problems arise when a nonconvex and nonsmooth objective function has to be optimized.[2] Deterministic gradient search methods might be comfortable to solve the optimization problem, as long as the considered measures have sufficient differentiability properties and the function is monotonically in- or decreasing. However, the proof of such differentiability properties can be rather difficult for random variables with a discrete probability distribution. In addition, the mean-variance-skewness function may have multiple feasible regions and multiple locally optimal points within each region. Thus, using derivative or gradient information in order to determine the direction in which the function is increasing (or decreasing) may be time-consuming. On a one-processor computer, a gradient algorithm would have to be started many times with a different set of decision variables ("brute-force"), as one might be frequently stuck in a local maxima and the situation at one possible solution gives very little information about where to look for a better solution. Many methods used to circumvent this problem have been restricted to goal programing or linear programing techniques. For example, [240] gave a goal programing procedure that performs portfolio selection based on competing and conflicting objectives by maximizing both expected return and skewness while minimizing the risk associated with the return (i.e., variance). Similarly, [253] provided a goal programing algorithm to solve a mean-variance-skewness model with the aid of the general Minkovski distance. Diverging from previous studies, [414] transformed the mean-variance-skewness model into a parametric linear programing problem by maximizing the skewness under given levels of mean and variance. Likewise, [265] also transformed the mean-variance-skewness model with transaction costs into a linear programing problem and verified its efficiency via a numerical example. However, the main disadvantage of these algorithms is that they generally converge slowly, if at all [224]. It is impractical and inefficient to exhaustively enumerate all of the possible solutions and pick the best one, even on a fast computer. Furthermore, most existing studies only present some numerical examples with artificial data.

---

[2] A set $S \subseteq R^n$ is convex if: $x', x'' \in S \Rightarrow \alpha x' + (1-\alpha)x'' \in S$, $0 \leq \alpha \leq 1$. A function $f(x)$ is convex on S if $x', x'' \in S$, $0 \leq \alpha \leq 1$, $f(\alpha x' + (1-\alpha)x'') \leq \alpha f(x') + (1-\alpha)f(x'')$. In other words, the line joining x' and x'' is never below the function value at that point. The fact that the objective function is nonconvex also implies that it is nonlinear, as it is a well known fact that a linear function is always convex (and concave).

Chapter 14
Model Implementation

14.1 Simulation/Optimization

In the absence of a tractable mathematical structure of Problem 13.1, we study the behavior of heuristics, "quick and dirty" algorithms which return feasible solutions that are not necessarily optimal. In particular, we apply a methodology called Simulation/Optimization which is a general expression for solving problems where one has to search for the settings of controllable decision variables that yield the maximum or minimum expected performance of a stochastic system as presented by a simulation model [146, 148]. For a compact picture of the field, see the reviews of the Winter Simulation Conference [147, 149].

Given the analogies between the TSP and the PTSP, it is reasonable to expect that, like in the TSP, a good heuristic for the problem may be obtained by the integration of

1. A solution construction algorithm generating candidate solutions
2. A local search algorithm, which tries to improve as much as possible the candidate solution

The sequence construction and improvement of a solution is repeated several times until a good solution or some other termination criterion is not satisfied. With respect to Fig. 14.1, this involves running a simulation for an initial set of values, analyzing the results, changing one or more values, rerunning the simulation, and repeating the process until a satisfactory (optimal) solution is found ([123], p. 13).

This process can be very tedious and time-consuming since not all of these values improve the objective function. The accuracy of the results therefore depends on the time limit granted for searching, the number of trials per simulation, the number of decision variables, and complexity of the objective function. It has been shown that even when there is no uncertainty, optimization can be very difficult if the number of decision variables is large, and little is known about the structure of the objective function. By definition, the main difference between the PTSP requiring simulation for estimating the objective function and the TSP with an exactly computable objective function is that, in the first-mentioned case, it is not possible to decide with

C. Ullrich, *Forecasting and Hedging in the Foreign Exchange Markets,* Lecture Notes
in Economics and Mathematical Systems 623, DOI: 10.1007/978-3-642-00495-7_14,
© Springer-Verlag Berlin Heidelberg 2009

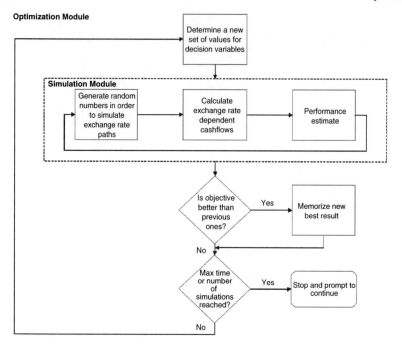

Fig. 14.1 Coordination of optimization and simulation

certainty whether a solution is better than another one. This can only be tested by statistical sampling, obtaining a correct comparison result only with a certain probability. The PTSP therefore adds an additional complication to the TSP because the performance of a particular candidate solution cannot be evaluated exactly, but instead must be estimated. It may be therefore not possible to conclusively determine if one candidate solution is better than another one which makes it difficult for optimization algorithms to move in improving directions. Theoretically, one can eliminate this complication by generating so many replications, or such long runs, at each candidate solution that the performance estimate has essentially no variance. In practice, however, this could mean that very few alternative candidate solutions will be explored due to the time required to simulate each one ([27], p. 488).

14.2 Simulation Model

14.2.1 Datasets

Given its launch on 1 January 1999, when the Euro became the currency for 11 member states of the European Union, the Euro exchange rate has a rather short history. As a result, it is not straightforward to scrutinize the evolution of the Euro over this period against longer term trends. In order to overcome this obstacle, it has become common practice to use a proxy measure for the Euro for the period before its actual existence. Either the DEM [80], which was de facto the anchor currency among the European currencies participating in the exchange rate mechanism, or a "synthetic" Euro exchange rate [309], i.e., a weighted average of the Euro legacy currencies has been commonly used. We decided for the first alternative and construct the EUR/USD series for the period from January 1975 to December 1998 by using the fixed EURP/DEM conversion rate agreed in 1998 (1 EUR = 1.95583 DEM). From January 1999 until March 2007, we use the regular EUR/USD exchange rate. The data we use are historical monthly averaged nominal exchange rate data as obtained from Bloomberg which total to 387 values.

14.2.2 Component 1: Equilibrium

According to economic theory, exchange rates are determined by relative price differentials in the long run. The relationship between price indices and the exchange rate is referred to as PPP which is one of the central doctrines in international economics (see Sect. 3.1). PPP is expressed by the price differential between two countries, which is given by the differential between the consumer price index (CPI) in the United States and the CPI of the Eurozone. Monthly CPI data was obtained from the OECD database for the same period of time as above. In order to adjust for the level of the exchange rate, the price differential was multiplied by the respective average spot rates over the observation period

$$PPP_t = \left(\frac{1}{T} \sum_{t=1}^{T} S_t \right) \frac{P_t^*}{P_t} \tag{14.1}$$

14.2.2.1 Visual Inspection

The EUR/USD exchange rate series and the equilibrium rate as determined by (14.1) is shown in Fig. 14.2.

Visual inspection seems to support that the EUR/USD real exchange rate oscillates around a mean in the long run. Between the January 1999 and October 2000, the EUR depreciated by around 20% in nominal effective terms. This reflected a significant weakening of the EUR against the currencies of major trading partners such as the US dollar (-26%). After reaching a trough in October 2000, the EUR bottomed out. Following a period of protracted weakness, the EUR moved in 2002 onto a steady recovery track to trade by March 2007 well above the levels observed at its launch, before reverting again somewhat in 2005 and also, well above equilibrium. The fact that rigorous statistical tests, such as the conventional unit root test, commonly fail to confirm such mean reverting properties [231, 359] should not prevent us from drawing inferences from PPP analysis. For example, if a unit root would

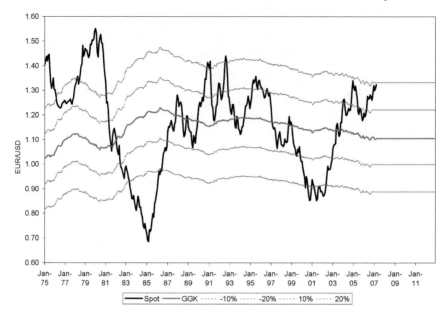

Fig. 14.2 EUR/USD equilibrium

indeed mimic the true data generating process of the real exchange rate, it would behave like a random walk without a systematic tendency to revert towards its PPP equilibrium. Figure 14.2 illustrates, however, that this is not the case. According to the first PPP puzzle, the so-called *disconnect puzzle* (see Sect. 3.1.3), exchange rates often get too far out of line with macroeconomic fundamentals in the medium term.

This is better illustrated by Fig. 14.3 which displays the nominal effective exchange rate as a percentage deviation from equilibrium.

We make three observations.

1. The EUR/USD exchange rate seems to be more persistent when it is in the proximity of the long-run mean, whereas reversion towards the mean happens more rapidly when the absolute size of PPP deviation is large. In fact, the further away the real exchange rate moves from PPP, the stronger the adjustment intensity becomes.
2. This behavior can be observed in both states of positive disequilibrium and negative disequilibria, which gives rise to a symmetric kind of adjustment behavior.
3. Within more than 30 years of data, the mean is crossed only 18 times which indicates a remarkable degree of persistence of the real exchange rate.

The observations are consistent with a strand of literature focusing on the existence of nonlinear dynamics in the real exchange rate, implying that the speed of mean reversion is state dependent [387, 403]. This literature suggests that the observed real exchange rates may be stationary around a trend, albeit persistent, and that it is very persistent in the neighborhood of PPP, while being mean-reverting at a faster speed when the deviation from PPP gets larger. Such models are called

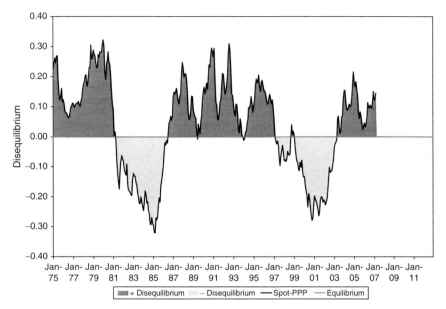

Fig. 14.3 EUR/USD disequilibrium

regime-switching models or (smooth) transition autoregressive ((S)TAR) models. References [387, 388, 403] have suggested a data-based modeling cycle for estimating STAR models. While this is certainly a rigorous procedure, it may lack economic intuition. We therefore propose a modeling procedure that is based on our observations and economic reasoning.

14.2.3 Component 2: Nonlinear Mean Reversion

14.2.3.1 Transition Function

Let $G[s_t; \gamma, c]$ be a continuous function (transition function) that determines the degree of mean reversion and has the following properties: $G[s_t; \gamma, c]$ is bounded between 0 and 1 and is itself governed by

- The transition variable s_t which is assumed to be a lagged endogenous variable [387], that is $s_t = r_{t-d}$ for certain integer delay parameter $d > 0$.[1]
- The parameter $\gamma > 0$ which effectively determines the speed of mean reversion.
- The locality parameter c which determines where the transition occurs. In case the exchange rate is located close to equilibrium, there is high uncertainty about

---

[1] The transition variable can also be represented by an exogenous variable, or a function of lagged endogenous variables, or a function of a linear time trend, as given by [264].

its future short-term course. Thus, it might be useful to determine a (close) corridor of exchange rate bandwidths that the future exchange rate is likely to pass through.

The regime that occurs at time t can be determined by the observable variable s_t and the associated value of $G[s_t; \gamma, c]$. Different choices for the transition function $G[s_t; \gamma, c]$ give rise to different types of regime-switching behavior. One popular choice includes the first-order logistic function $G[s_t; \gamma, c] = (1 + \exp\{-\gamma(s_t - c)\})^{-1}$ which has been used for modeling asymmetric cycles of the underlying, for example, business cycle asymmetry to distinguish expansions and recessions [389, 390]. However, in the case of exchange rates, this may not make sense. Our visual inspections and economic reasoning rather imply the existence of a symmetric band around the equilibrium rate in which there is no tendency of the real exchange rate to revert to its equilibrium value (inner regime). Outside this band (outer regime), commodity arbitrage becomes profitable, which forces the real exchange rate back towards the band. If regime-switching of this form has to be captured, it appears more appropriate to specify the transition function such that the regimes are associated with small and large absolute values of s_t. This can be achieved by using the exponential (E)STAR function $G[s_t; \gamma, c] = 1 - \exp\{-\gamma(s_t - c)^2\}$ according to [174]. The ESTAR function has been successfully applied to real exchange rates by ([292, 386]) and to real effective exchange rates by [355]. However, according to [403], a drawback of the exponential function is that for either $\gamma \to 0$ or $\gamma \to \infty$ the function collapses to a constant (equal to 0 and 1, respectively). Hence, the model becomes linear in both cases and the ESTAR model does not nest a self-exciting (SE)TAR model as a special case, as would be given by the second-order logistic function

$$G[s_t; \gamma, c] = (1 + \exp\{-\gamma(s_t - c_1)(s_t - c_2)\})^{-1}, \; c_1 \leq c_2, \; \gamma \geq 0 \qquad (14.2)$$

where now $c = (c_1, c_2)'$, as proposed by [205]. The properties of the second-order logistic function are depicted in Fig. 14.4.

In this case, if $\gamma \to 0$, the model becomes linear, whereas if $\gamma \to \infty$ and $c_1 \neq c_2$, the function $G[s_t; \gamma, c]$ is equal to 1 for $s_t < c_1$ and $s_t < c_2$ and equal to 0 in between. Hence, the STAR model with this particular transition function nests a restricted three-regime SETAR model, where the restriction is that the outer regimes are identical. Note that for moderate values of γ, the minimum value of the second-order logistic function, attained for $s_t = (c_1 + c_2)/2$ remains between 0 and 0.5, unless $\gamma \to \infty$. In the latter case, the minimum value does equal zero. This has to be kept in mind when interpreting estimates from models with this particular transition function.

14.2.3.2 Illustration and Theoretical Rational

Let us further illustrate the concept of mean reversion by translating the logistic transition function of second order into the classical price chart. The future spot rate

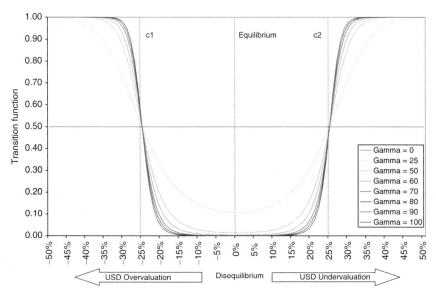

Fig. 14.4 LSTAR2 transition functions for varying degrees of γ and fixed c

describes the deterministic sum of the PPP level at $t = 0$ and the future real exchange rate $\hat{r}_{t+1}$

$$\hat{s}_{t+1} = PPP_{t=0} + \hat{r}_{t+1} \tag{14.3}$$

where

$$\hat{r}_{t+1} = \begin{cases} r_t & : & c_1 \leq s_t \leq c_2 \\ r_t \times G[s_{t=0}; \gamma, c] & : & \text{else} \end{cases} \tag{14.4}$$

Hence, the future real exchange rate is dependent on the value of the transition function $G[s_t; \gamma, c]$ if s_t lies in the outer regime, i.e., $s_t < c_1$ or $s_t > c_1$.

Figure 14.5 illustrates the smooth adjustment of future spot to the level of PPP at March 2005 according to (14.3) with $PPP_{t=0} = 1.1250$ and $G[s_{t=0} = 1.3184, \gamma = 50, (c_1, c_2) = (0,0)]$. Equilibrium theory states that if exchange rates get too far out of line with macroeconomic fundamentals, "gravitational forces" must bring the system back to equilibrium. These forces lead to nonlinearities in the adjustment of exchange rates which can be justified from both a goods and a financial market perspective. As international goods arbitrage involves transaction costs, such as costs of transportation and storage of goods, arbitrage sets in once the real exchange rate protractedly moves outside certain limits. Consequently, commodity arbitrage does not compensate for the costs involved in the necessary transactions for small deviations from the equilibrium real exchange rate. From an asset market perspective, differing trading strategies could play a role with *chartists* dominating market dynamics when the exchange rate is close to some perceived equilibrium, while *fundamentalists* being at play once the exchange rate is increasingly misaligned.

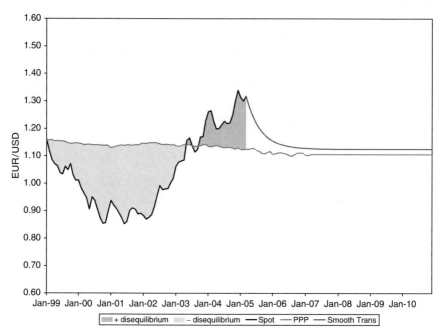

Fig. 14.5 Nonlinear mean reversion

14.2.4 Component 3: Gaussian Random Walk

14.2.4.1 Definition

According to the second PPP puzzle, the so called *excess volatility puzzle* (see Sect. 3.1.3), the enormous short-term volatility of real exchange rates is contradicting to the extremely slow rate of adjustment. As we observed, despite the mean-reverting behavior, the real exchange rate also indicates a remarkable degree of persistence. We therefore assume that in the short term volatility becomes increasingly important. In fact, the adjustment process can be overshadowed/dominated by market volatility. This phenomenon is modeled by a simple one-dimensional random walk which is given by the AR(1) process

$$\hat{s}_{t+1} = s_t + \hat{\varepsilon}_{t+1} \tag{14.5}$$

where $\hat{\varepsilon}_{t+1} \sim NID(0, \sigma^2)$ with constant mean and finite variance.

The normal distribution is based on the central limit theorem, which describes a tendency as the number of observations becomes infinitely large. Typically, this assumption is not made because it allows to model financial markets more accurately than competing frameworks, but because it is transparent and tractable. For example, the assumption of normally distributed returns need not imply that we believe that returns have been normal in the past, or expect them to be normal in the future. We may simply believe that forecasts based on the normal distribution have not done significantly worse than those based on other more sophisticated models. Or, even if we believe that other models have done better in the past, we may be unable to predict that they will do better in the future. In addition, it is clear that our way of attempting to model the market is first and foremost determined by the sampling frequency of interest. In contrast to many financial firms, industrial firms naturally view their financial market exposures over monthly, or quarterly periods of time. As a consequence, there is no need to sample the market on a daily or even high-frequency basis and thus there is no need for modeling excess kurtosis, i.e., fat tails. As the news flow to markets causes less variation in returns when comparing longer periods, extreme values are averaged out and therefore, normality tends to become more pronounced as the sampling frequency of price changes decreases ([375], pp. 145–165).

14.2.4.2 Simulation of a One-Dimensional Random Walk

Exchange rate returns can be computer-simulated by randomly (independently) drawing from historically observed returns (historical simulation), or drawing randomly (independently) percentiles from a certain probability distribution (Monte Carlo simulation). "Independently" means that every simulated exchange rate return ε_t is independent of ε_{t-1}. We opt for the simple Monte Carlo method [202], in which the range of probable values for an uncertain input parameter is divided into ordered segments of equal probability. The values are sampled in such a way as to generate samples from all the ranges of possible values, producing information about the extremes of the output probability distributions. Compared to variance-reduction sampling techniques, such as the Latin Hypercube method, which reduces the number of solutions (and may therefore neglect the likelihood of extremes) in order to decrease computation time, the simple method is found to be more appropriate in our context, that is for financial problems involving the assessment of risk. Based on today's value of the spot rate s_0, randomly generated future exchange rate returns are linked to paths of length T.

The exchange rate s_t is derived from the exchange rate at s_{t-1} to which the simulated change ε_t was added, where ε_t can be either positive or negative. According to this principle exchange rate paths of any length can be simulated. The procedure can be summarized in four steps as follows:

1. Determine s_0 (which is the spot rate in our case), I (which is the number of scenarios), T (which denotes the simulation horizon)
2. Simulation: Set $i = 1$. Generate the ith scenario $\varepsilon_{i,t}$ for every period $t = 1, 2, \ldots, T$ within the simulation horizon
3. Path Linking: $s_t = s_0 + \varepsilon_{i,1} + \varepsilon_{i,2} + \ldots + \varepsilon_{i,T}$
4. Iteration: set $i = i + 1$; $\varepsilon_{i,t}, i = 1, 2, \ldots, I; t = 1, 2, \ldots, T$ until $i = I$.

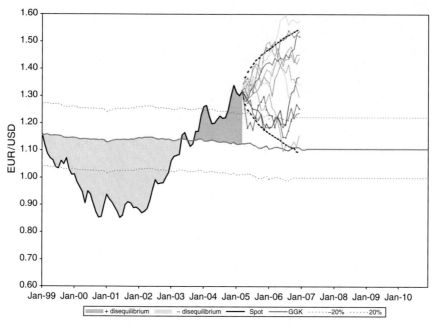

Fig. 14.6 Random walk without drift

Figure 14.6 illustrates 10 out of 10,000 randomly generated, driftless, exchange rate scenarios starting at a spot rate of EUR/USD = 1.3184 as per March 2005 and an estimated standard deviation of $\sigma = 0.025$. The dotted bold black lines represent the 5% and 95%, i.e., according to our assumptions there is a 90% probability that the exchange rate will move within this bandwidth.

14.2.5 Aggregation of Components

14.2.5.1 Gaussian Random Walk with Nonlinear Drift

The aggregation of the three single components results in a random walk model with nonlinear drift. Let $\Delta_{t+1} = r_{t+1} - r_t$ be the difference of the real exchange rate from the next state to the state of today and let r_{t+1} be defined according to (14.4). Then (14.5) can be written as

$$s_{t+1} = PPP_{t=0} + r_{t+1} = s_t + \Delta_{t+1} \tag{14.6}$$

From (14.6) and (14.5) we obtain

$$s_{t+1} = s_t + \Delta_{t+1} + \varepsilon_{t+1}, \ t = 0, 1, 2 \dots, T \tag{14.7}$$

where $\varepsilon_{t+1} \sim NID(\Delta_{t+1}, \sigma)$.

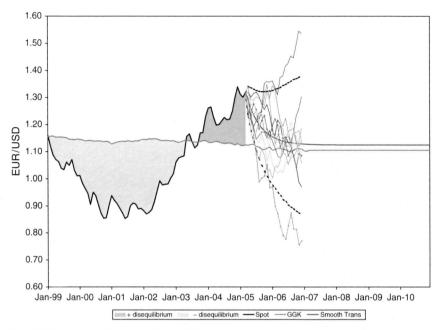

Fig. 14.7 Random walk with nonlinear drift

14.2.5.2 Illustration

The behavior of the model is illustrated by Fig. 14.7 and illustrates 10 out of 10,000 randomly generated, exchange rate scenarios starting at a spot rate of EUR/USD = 1.3184 as per March 2005 and an estimated standard deviation of $\sigma = 0.025$. The drift is modeled by smooth adjustment of future spot to the level of PPP as per March 2005 with $PPP_{t=0} = 1.1250$ and $G[s_{t=0} = 1.3184, \gamma = 50, (c_1, c_2) = (0,0)]$. Accordingly, while the Euro real exchange rate can be well approximated by a random walk if PPP deviations are small, in periods of significant deviations, gravitational forces are set to take root and bring the exchange rate back towards its long-term trend. Exchange rate scenarios are PPP-reverting as shown by the 5% and 95% bandwidths. Still, there may be scenarios where PPP adjustment is overshadowed by market volatility and no direct adjustment can be recognized. This behavior is realistic since it matches the second PPP puzzle.

14.2.6 Calibration of Parameters

14.2.6.1 Estimation of γ and c

We recommend to estimate the slope parameter γ and the location parameter c of (14.4) in three steps as follows:

1. Given the current market circumstances, identify c subjectively
2. Estimate γ by OLS, conditional on c
3. Undertake maximum likelihood estimation of the selected nonlinear model from Step 2, including estimation of γ and c

We found this procedure to work well in practice. Recall that the locality parameter c determines where the transition occurs. From an asset market perspective, we recommend to estimate c subjectively, rather than determining it by a purely data driven approach [387, 403]. If "chartists" dominate market dynamics, then there usually exists something like a "perceived equilibrium," i.e., a critical short term threshold. This may be hard to determine based on historical exchange rate data since it depends on market expectations. Instead, one might use short-term information about technical equilibria as obtained from banks and brokers. Alternatively, companies might also use the interest rate differential as given by the forward rate, as an indicator for short-term equilibrium value. For instance, one could perceive technical equilibrium to be 5% above/(below) fundamental equilibrium, i.e., $c = (c_1, c_2) = ((1 - \delta)PPP_{t=0}, (1 + \delta)PPP_{t=0})$ with $\delta = 0.05$. This would imply that, in case the exchange rate approaches c, there will be no further adjustment towards PPP since the model collapses to a Gaussian random walk without drift.

Given c, we estimated γ by OLS, i.e.,

$$\min Z = \sum_{t=1}^{T} \sum_{i=1}^{I} (r_i - \hat{r}_i)^2 \tag{14.8}$$

with $\hat{r}_i$ according to (14.4) and $t = 1, 2, \ldots, T$ describes the backtesting period. The index $i = 1, 2, \ldots, I$ describes the forecasting period for γ which was set to $I = 36$ months. OLS was performed with the Microsoft Excel 2000's standard solver ([150]) in order to derive optimal values for γ.

14.2.6.2 Estimation of σ

The standard deviation of $\hat{\varepsilon}_{t+1}$ from (14.7) was estimated by taking the square root of an unbiased estimate of the empirical variance

$$\hat{\sigma}_t = \sqrt{\frac{1}{t-1} \sum_{i=t-23}^{t} (s_i - \bar{s})^2} \tag{14.9}$$

where $\bar{s}$ is the mean of EUR/USD first differences over the last 24 months.

14.3 Optimization Model

14.3.1 Solution Construction Algorithm

We propose the use of a metaheuristic combinatorial search algorithm. The motivation is twofold. First, we have shown that our problem is *NP*-hard and cannot be solved in polynomial time which directly suggests that solutions can only be approximated. Second, the problem is nonconvex and nonsmooth which makes it difficult to find an optimal solution since one cannot simply follow the gradient of the objective function.

Metaheuristics are modern heuristics which have been first introduced in [161, 162], and have proven to be a very exciting and practical development in approximate optimization techniques. They have had widespread successes in attacking a variety of difficult combinatorial optimization problems that arise in many practical areas. Osman and Laporte [316] conducted a comprehensive list of 1,380 references on the theory and application of metaheuristics. A more recent survey on metaheuristics in stochastic combinatorial optimization has been conducted by [36]. Until today, there is no commonly accepted definition, although researchers have made many attempts (see, for instance, [46, 316]). In short, a metaheuristic refers to a master strategy that guides and modifies other heuristic methods to produce solutions beyond those that are normally generated in a quest for local optimality. In order to do so, metaheuristics incorporate a dynamic balance between intensification and diversification. The term intensification refers to the exploitation of the accumulated search experience, whereas the term diversification, generally refers to the exploration of the search space.[2] Some researchers consider these two requirements to be naturally conflicting [28], others to be interplaying [163]. Regarding the latter, the distinction between intensification and diversification is often interpreted with respect to the temporal horizon of the search.

> In some instances we may conceive of intensification as having the function of an intermediate term strategy, while diversification applies to considerations that emerge in the longer run. [163]

Short-term search strategies can therefore be seen as the iterative application of tactics with a strong intensification character. When the horizon is enlarged, usually strategies referring to some sort of diversification come into play. Either way, a balance between intensification and diversification must be found, on the one hand to quickly identify regions in the search space with high quality solutions, and on the other hand not to waste too much time in regions of the search space which are either already explored or which do not provide high quality solutions. The balance between intensification and diversification might be quite different not only across different metaheuristics, but also for different components, which are generally understood as operators, actions, or strategies of a specific metaheuristic algorithm.

---

[2] Note that diversification is not the same as randomization. Whereas the goal of diversification is to produce solutions that differ from each other in significant ways, and that yield productive alternatives in the context of the problem considered, the goal of randomization, is to produce solutions that may differ from each other in any way, as long as the differences are entirely "unsystematic."

14.3.2 Scatter Search and Path Relinking

14.3.2.1 Pseudocode Template

We use a procedure that is based on the scatter search methodology, but incorporates innovative mechanisms to exploit the knowledge of the problem and to create a trade-off between intensification and diversification for an efficient search. The seminal ideas of scatter search originated in the late 1960s. A first description of scatter search appeared in [161]. The modern version of the method is described in [239]. Scatter search and its generalized form, called path relinking [166, 167], have proven to be highly successful for a variety of known problems, such as vehicle routing [18, 337, 338], tree problems [68, 423], mixed integer programing [165], or financial product design [81].

Scatter search is a search strategy that operates on a set of solutions, the reference set, by generating linear combinations of the reference solutions to create new ones. The resulting solutions are called trial solutions. These trial solutions may be infeasible solutions and are therefore modified by means of a repair procedure that transforms them into feasible solutions. A heuristic improvement mechanism is then applied in order to try to improve the set of trial solutions. These improved solutions form the set of dispersed solutions. The new set of reference solutions that will be used in the next iteration is selected from the current set of reference solutions and the newly created set of dispersed solutions. The coupling of heuristic improvement with solution combination strategies, as is the case in scatter search, has inaugurated the term memetic algorithms (see, e.g., [300]). The Pseudocode template for scatter search, according to [163], has served as the main reference for most of the scatter search implementation and can be stated as according to Fig. 14.8. The template is divided into two phases: an initial phase that generates a first set of dispersed solutions, and a scatter search or path relinking phase that seeks to improve the initial solution by iteratively searching through the trajectory.

The six components processes may be sketched in basic outline as follows:

```
1:  Initial Phase:
2:        Seed Generation()
3:  Repeat
4:        Diversification Generator()
5:        Improvement()
6:        ReferenceSetUpdate()
7:  Until the reference is of cardinality n
8:  Scatter Search/Path Relinking Phase:
9:  Repeat
10:       Subset Generation()
11:       Solution Combination()   //Path Relinking
12:       Improvement()
13:       ReferenceSetUpdate   //Tabu List
14:  Until termination criteria met
```

Fig. 14.8 Scatter Search pseudocode template

1. Seed: create one or more seed solutions, i.e., arbitrary trial solutions used to initiate the remainder of the method.
2. Diversification Generation Method: generate a collection of diverse trial solutions, using one or more seed solutions as an input. Check feasibility of trial solutions and repair solution if it is infeasible.
3. Improvement Method: transform a trial solution into one or more enhanced trial solutions.
4. Reference Set Update Method: build and maintain a reference set consisting of the b best solutions found, organized to provide efficient accessing by other parts of the solution procedure. Several alternative criteria may be used to add solutions to the reference set and delete solutions from the reference set.
5. Subset Generation Method: operate on the reference set to produce a subset of its solutions as a basis for creating combined solutions. The most common subset generation method is to generate all pairs of reference solutions.
6. Solution Combination Method: transform a given subset of solutions produced by the Subset Generation Method into one or more combined solutions.

The interconnection between above methods is described in Pseudocode form by Fig. 14.9 [163] and will be explained in more detail in the next paragraphs. The scatter search methodology is very flexible since each of its above methods can be implemented in a variety of ways and degrees of sophistication. The advanced features of scatter search are therefore related to the way these methods are implemented. That is, the sophistication comes from the implementation of she scatter search methods instead of the decision to include or exclude certain elements (as in the case of tabu search or other metaheuristics).

14.3.2.2 Implementation

Diversification Generation Method

The diversification generation method is used only once to build a large set of $PSize$ different solutions P at the beginning of the search, where $PSize$ is 10 times the size of $RefSet$, and is never employed again. An initial reference set of solutions ($RefSet$) is created which consists of $b = PSize/10$ distinct and maximally diverse solutions. Consider the midpoint of the initial population

$$x_i = l_i + \frac{(u_i - l_i)}{2}, \ i = 1, \dots, n \tag{14.10}$$

where $L = \{l_i : i = 1, \dots, n\}$ is the set of lower bound values and $U = \{u_i : i = 1, \dots, n\}$ is the set of upper bound values for all candidate solutions $x_i \in X$. Additional points are generated with the goal of creating a diverse population. A population is considered diverse if its elements are significantly different from one another. An Euclidean distance measure $d(x, y)$ is used to determine how "close" a potential new point x, i.e., a candidate solution, is from the points y already in the population

1. Start with $P = \emptyset$.

    - Use the diversification generation method to construct a solution y
    - Apply the improvement method to y. Let x be the final solution
    - If $x \notin P$ then add x to P (i.e., $P = P \cup x$), otherwise, discard x
    - Repeat step until $|P| = PSize$. Build $RefSet = \{x^1, \ldots, x^b\}$ with b diverse solutions in P

2. Evaluate the solutions in $RefSet$ and order them according to their objective function value such that x^1 is the best solution and x^b the worst. Make $NewSolutions = $TRUE.

    while ($NewSolutions$) **do**

    3. Generate $NewSubsets$, which consists of all pairs of solutions in $RefSet$ that include at least one new solution. Make $NewSolutions = $FALSE.

    while ($NewSubsets \neq \emptyset$) **do**

        4. Select the next subset s (i.e., the next pair (x', x'')) in $NewSubsets$.

        5. Apply the Solution Combination Method to s in order to obtain one or more new solutions y.

        //Path Relinking

        Produce the sequence $x' = x(1), x(2), \ldots, x(r) = x''$

        for $i = 1$ **to** $i < r/NumImp$ **do**

        6. Apply the Improvement Method to $x(NumImp \times i)$.

        end for

        7. Apply the Relinking Method to produce the sequence $x'' = y(1), y(2), \ldots, y(s) = x'$

            for $i = 1$ **to** $i < s/NumImp$ **do**

        8. Apply the Improvement Method to $y(NumImp \times i)$ and obtain x.

        end for

        for (each generated solution x)

            if (x is not in $RefSet$ and $f(x) < f(x^b)$) **then**

            9. Make $x^b = x$ and reorder $RefSet$.

            10. Make $NewSolutions = $TRUE.

            end if

        end for

        11. Delete $s = (x', x'')$ from $NewSubsets$.

    end while

    end while

Fig. 14.9 Scatter Search/Path Relinking pseudocode

(in *RefSet*), in order to decide whether the point is included or discarded. That is, for each candidate solution x in $P - RefSet$ and reference set solution y in *RefSet*, a measure of distance or dissimilarity $d(x,y)$ is calculated. The candidate solution is selected which maximizes

$$d_{min}(x) = min_{y \in RefSet}\{d(x,y)\} \qquad (14.11)$$

Since linear constraints are imposed on a solution x, every reference point x is subject to a feasibility test before it is evaluated, i.e., before the simulation model is run to determine the value of $f(x)$. The feasibility test consists of checking whether the linear constraints imposed are satisfied. An infeasible point x is made feasible by formulating and solving a linear programing (LP) problem with the goal of finding a feasible x^* that minimizes the absolute deviation between x and x^*.

$$\min d^- + d^+$$
$$\text{subject to } AX^* \leq B \qquad (14.12)$$
$$X - X^* + d^- + d^+ = 0$$
$$L \leq X^* \leq U$$

where d^- and d^+ are negative and positive deviations from the feasible point x^* to the infeasible reference point x.

Improvement Method

Once the population is generated, the procedure iterates in search of improved outcomes. In each iteration two reference points are selected to create four offsprings. Let the parent-reference points be x_1 and x_2, then the offspring x_3 to x_6 are found as follows:

$$x_3 = x_1 + d$$
$$x_4 = x_1 - d \qquad (14.13)$$
$$x_5 = x_2 + d$$
$$x_6 = x_2 - d$$

where $d = (x_1 - x_2)/3$. The selection of x_1 and x_2 is dependent on the objective function values $f(x_1)$ and $f(x_2)$, as well as the search history. An advanced form of Tabu Search (see e.g., [164]) is superimposed to control the composition of reference points at each stage. In its simplest manifestations, adaptive memory is exploited to prohibit the search from reinvestigating solutions that have already been evaluated. However, the use of memory of the applied algorithm is much more advanced and calls upon memory functions that encourage search diversification and intensification in order to allow the search to escape from locally optimal solutions and possibly find a globally optimal solution. For instance, in the course of searching for a global optimum, the population may contain many reference points with similar characteristics. That is, in the process of generating offspring from a mixture of high-quality reference points and ordinary reference point members of the current population, the diversity of the population may tend to decrease and the likelihood increases that the system gets stuck in a local optimum. A strategy that remedies this situation considers the creation of a new population by a restarting mechanism which intends to create a population that is a blend of high quality points found in earlier explorations (referred to as elite points) complemented with points generated in the same way as during the initialization phase. The restarting mechanism, therefore, injects diversity through newly generated points and pre-serves quality through the inclusion of elite points. Diversity is achieved by means of an age strategy, a form of long-term memory, which assigns to each candidate solution of a population a measure of "attractiveness." Since some of the points in the initial population may have poor objective function values with a high prob-ability, they may never be chosen to play the role of a parent and would remain in the population until restarting. To additionally diversify the search, the system increases the attractiveness of these unused points over time. Thus, at the start of the search process, all the reference points x in a population of size p have an age of zero. At the end of the first iteration, there will be $p-1$ reference points from the original population and one new offspring. The ages of the $p-1$ reference points are set to one and that of the new offspring is set to zero. The same logic is followed in subsequent iterations and therefore, the age of every reference point increases by one in each iteration except for the age of the new population member which is initialized to zero. Each reference point in the population has an associated age and an objective function value. These two values are used to define a function of attractiveness that makes an old high-quality point the most attractive. Low-quality points become more attractive as their age increases.

In order to speed up the system's search engine, a neural network is used to as a prediction model to help the system accelerate the search by avoiding the need for evaluating $f(x)$ for a newly created reference point x, in situations where the $f(x)$ value can be predicted to be of low quality [239]. During the search, the neural network is trained by historical values of x and $f_\alpha(x)$ which are collected for a number of past iterations, where α denotes a percentage of the total number of simulation trials. The predicted value of the objective function for x as determined by the neural network is therefore given by $\hat{f}(x, f_\alpha(x))$. During each training round, an error value is calculated which reflects the accuracy of the network as a prediction model. That is, if the network is used to predict $f(x)$ for a newly created x, then the error indicates how good the prediction is expected to be. The error term is calculated by computing the differences between the known $f(x)$ and the predicted $\hat{f}(x, f_\alpha(x))$ objective function values. The training continues until the error falls within a specified maximum value.

Reference Set Update Method

The *RefSet* Update method accompanies each application of the improvement method. The update operation is to maintain a record of the b best solutions found.

Let *bNow* denote the current number of solutions in *RefSet*. Initially, *bNow* begins at zero, and is increased each time a new solution is added to *RefSet*, until reaching the value *bMax*. At each step, *RefSet* stores the best solutions in an array $x[i]$, $i = 1$ to *bNow*. An associated location array $loc(i)$, $i = 1$ to *bNow*, indicates the ranking of the solutions. That is, $x[loc(1)]$ (the x vector stored in location $loc(1)$) is the best solution, $x[loc(2)]$ is the next best solution, and so on. A solution $x = x'$ is not permitted to be recorded, if it duplicates another already in *RefSet*. The search terminates when no new solutions are admitted to *RefSet*.

Subset Generation Method

Another important feature relates to the strategy for selecting particular subsets of solutions to combine. According to Step 3 in Fig. 14.9, *NewSubsets* is constructed by systematically including all those pairs of *RefSet* elements that contain at least one new solution. This means that the procedure does not allow for two solutions to be subjected to the solution combination method more than once. We consider subsets of size 2, and specify that the cardinality of *NewSubsets* corresponding to the initial reference set is given by $(b^2 - b)/2$, which accounts for all pairs of solutions in *RefSet*.

Solution Combination (Path Relinking) Method

After *NewSubsets* is constructed and NewSolutions is switched to FALSE, the algorithm chooses two elements of the reference set out of *NewSubsets* in lexicographical order and the solution combination method is applied to all pairs of solutions in the *RefSet* in order to generate one or more solutions in Step 5. Solutions are combined in a linear way through path relinking, an approach that was originally suggested to integrate intensification and diversification strategies in the context of tabu search ([164]). Here, path relinking extends the combination mechanisms of scatter search. Instead of directly producing a new solution when combining two or more original solutions, path relinking generates paths between and beyond the selected solutions in the neighborhood space. The desired paths are generated by selecting moves that perform the following role: upon starting from an initiating solution x', the moves must progressively contain attributes from a guiding solution x'' (or reduce the distance between attributes of the initiating and guiding solutions). Consider the creation of paths that join two selected reference solutions x' and x'', restricting attention to the part of the path that lies between the solutions, producing a solution sequence $x' = x(1), x(2), \ldots, x(r) = x''$. The combinations are based on the following three types:

- $x = x' - d$
- $x = x' + d$
- $x = x'' - d$

where $d = r \times (x'' - x')/2$, and r is a random number in the range (0,1).

Combinations beyond the segment $[x',x'']$, "outside" of x'' (i.e., nonconvex combinations), are generated by three strategies:

- Strategy 1: Computes the maximum value of r that yields a feasible solution x when considering both the bounds and the constraints in the model.
- Strategy 2: Considers the fact that variables may hit bounds before leaving the feasible region relative to other constraints. The first departure variable from the feasible region may happen because some variable hits a bound, which is followed by the others, before any of the linear constraints is violated. In such a case, the departing variable is fixed at its bound when it hits it, and the exploration continues with this variable held constant. This is done with each variable encountering a bound before other constraints are violated. The process finishes when the boundary defined by the other constraints is reached.
- Strategy 3: Considers that the exploration hits a boundary that may be defined by either bounds or any of the linear constraints. When this happens, one or more constraints are binding and the corresponding r-value cannot be increased without causing the violation of at least one constraint. At this point, OCL chooses a variable to make a substitution that geometrically corresponds to a projection that makes the search continue on a line that has the same direction, relative to the constraint that was reached, as it did upon approaching the hyperplane defined by the constraint. The process continues until the last unfixed variable hits a constraint. At this point, the value of all the previously fixed variables is computed.

Each of these three boundary strategies generates a boundary solution x^b outside x''. A fourth solution is being generated in the midpoint between x^b and x''. Interchanging the role of x' and x'' gives the extension outside the "other end" of the line segment.

The Improvement Method is applied every *NumImp* steps of the relinking process in each path (Steps 6 and 8). Each of the generated solutions, in each path, including also those obtained after the application of the Improvement Method, is checked to see whether it improves upon the worst solution currently in *RefSet*. If so, the new solution replaces the worst and *RefSet* is reordered in Step 6. The *NewSolutions* flag is switched to TRUE and the pair (x',x'') that was just combined is deleted from *NewSubsets* in Step 11. The number of solutions created from the linear combination of two reference solutions depends on the quality of the solutions being combined. Specifically, when the best two reference solutions are combined, they generate up to five new solutions, while when the worst two solutions are combined they generate only one. In the course of searching for a global optimum, the combination method may not be able to generate solutions of enough quality to become members of the reference set. If the reference set does not change and all the combinations of solutions have been explored, a diversification step is triggered. This step consists of rebuilding the reference set to create a balance between solution quality and diversity. To preserve quality, a small set of the best (elite) solutions in the current reference set is used to seed the new reference set. The remaining solutions are eliminated from the reference set. Then, the diversification generation method is used to repopulate the reference set with solutions that are diverse with respect to the elite set. This reference set is used as the starting point for a new round of combinations.

The structured combinations produced by scatter search are designed with the goal of creating weighted centers of selected sub-regions. These include nonconvex combinations that project new centers into regions that are external to the original reference solutions. The dispersion patterns created by such centers and their external projections are therefore based on both current and past evaluations of inputs. The procedure carries out a nonmonotonic search, where the successively generated inputs produce varying evaluations, not all of them improving, but which over time provide a highly efficient trajectory to the best solutions. The process continues until an appropriate termination criterion is satisfied, such as a particular amount of iterations or an amount of time to be devoted to the search.

Chapter 15
Simulation/Optimization Experiments

15.1 Practical Motivation

A situation is presented, where a manufacturing company, located in Eurozone, sells its goods via foreign subsidiaries to the end-customer in the United States.

Generally, a strong foreign currency (weak EUR) is considered to be beneficial for the company because of a larger purchasing power for the customer abroad. Since one unit of foreign currency is worth more units of home currency, Eurozone manufactured goods – if prices remain constant – are cheaper which in theory has a positive effect on international sales. Foreign exchange risk arises at the United States subsidiary, which on the 15th of every month needs to exchange its USD turnover into EUR in order to meet its liabilities with the Eurozone parent company. Since it is not clear what the spot exchange rates for the given currencies will be on the future transaction dates, the subsidiary is exposed to foreign exchange transaction risk. In this case, risk consists in an appreciating EUR against foreign currency. For the US subsidiary, a stronger EUR is associated with higher cost of goods and a lower turnover at period end as a consequence. The described situation is illustrated in Fig. 15.1.

15.2 Model Backtesting

15.2.1 Overview

For the purpose of model validation, historical data backtesting was carried out via a dynamic rolling window approach. We split the historical data into two parts. Part 1 comprises 168 monthly observations (January 1985 to December 1998) and Part 2 comprises 72 months (January 1999 to December 2004) of the total 96 months (January 1999 to December 2006). We do not conduct backtesting of the decision

C. Ullrich, *Forecasting and Hedging in the Foreign Exchange Markets,* Lecture Notes in Economics and Mathematical Systems 623, DOI: 10.1007/978-3-642-00495-7_15, © Springer-Verlag Berlin Heidelberg 2009

Fig. 15.1 Case study for
hedging EUR/USD

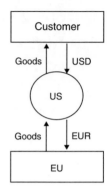

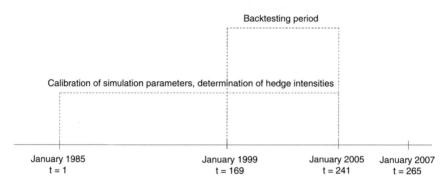

Fig. 15.2 Backtesting procedure

model over the remaining 24 months of the historical sample (January 2005 to
December 2006) because when solving the model we need 24 months of future
actual realized exchange rates.

Thus, the last roll of the decision model should be 24 months before the end of
the historical sample of exchange rates. In Fig. 15.2 we explain this using a time
line where January 1999 and January 2005 indicate the months, when backtesting
starts and ends, respectively. The experimental set up is progressively described in
the following.

15.2.2 Data Inputs and Parameters

15.2.2.1 Exposures

Exposures are assumed to be deterministic and set to 1 at any time. This allows us
to focus solely on exchange rate effects, and eases interpretation of results. If we
consider the fact that in a multinational firm, exposure forecasting is not a typical

risk management discipline but rather a sales planning task that is usually taken as given in practice, assuming deterministic exposures is not an unrealistic assumption either.

15.2.2.2 Hedge Intensity

The intensity of a hedge is a fraction of the total estimated exposure within a certain period of time.

It is formally captured by the variable z in Problem 13.1. The hedge intensity is calculated according to the procedure described in Fig. 15.3. In a PPP-reverting world, the cheaper it is to buy EUR according to the equilibrium rate, the higher should be the intensity of a hedge. This logic is captured by the simple Pseudocode as written in Fig. 15.3. For instance let min $= 0.6850$, max $= 1.5508$ and $I = 24$. Then the grid range is evenly divided into 24 slots of size $b = 0.0361$. If current spot is 0.9815, then $t^* = 15$ is returned, suggesting that the exposures for the next 15 months should be hedged. Note that according to Problem 13.1, z is nonnegative and hence, imposes a speculation constraint. The minimum intensity must therefore be $t^* = 0$ months, in case the spot exchange rate exceeds the historical maximum and is thus, considered to be very unfavorable. The maximum intensity is $t^* = 24$ months in case the spot exchange rate is lower than the historical minimum.

Figure 15.4 gives the results of the routine as applied on the period from January 1985 to December 2004. Hedge intensities are depicted by bar charts which refer to the y-axis on the right-hand side. The chart demonstrates that according to the proposed procedure, hedge intensities are perfectly correlated with the degree of disequilibrium. Hence, this approach of determining hedge intensities is consistent to the broader logic of foreign exchange hedging in a PPP-reverting context.

1. Determine grid range:

   - from January 1985 until 'current date': identify minimum spot rate (*minspot*) and maximum spot rate (*maxspot*)

2. Set maximum hedge intensity to $I = 24$
3. Calculate grid slot size $b = (maxspot - minspot)/I$
4. Set up grid:

   - define array of length $I + 1$
   - **for** $0 \leq i \leq I$: calculate $g[i] = minspot + b \times i$

5. Determine hedge intensity for current month:

   - Check in which grid slot current spot falls
   - **for** $0 \leq i \leq I$: **if** current spot $\leq g[i]$, return hedge intensity i

Fig. 15.3 Pseudocode template for calculation of hedge intensity

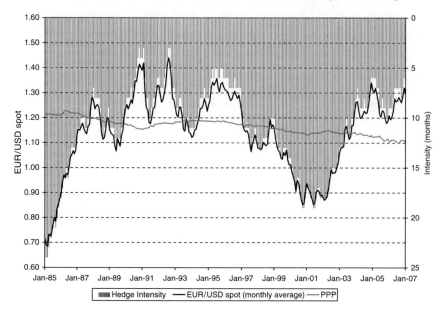

Fig. 15.4 Historical hedge intensities

15.2.2.3 Exchange Rate Data

For the 15th of every month in the backtesting period, historical spot exchange rates, forward exchange rates, strike prices and option premia for the following 24 months were obtained from Bloomberg. According to CIP, the no-arbitrage price of the forward contract or the forward price is determined by the spot rate today and the interest rate differential between the two currencies for the relevant maturity, also known as the forward premium. Figure 15.5 depicts EUR/USD forward premia, which correspond to the interest rate differentials between Eurozone and the United States for maturities from 1 to 24 months [43]. The forward rate at a certain date for a particular maturity is simply obtained by adding the respective forward premium on this date's spot rate.

Without loss of generality, it is assumed that the firm uses a single type of European option. Due to the existence of transaction costs, it is never optimal to combine options with different strike prices in this setting. In particular, a forward contract can be replicated with two options by put-call parity [200]. However, the synthetic forward produced in this way is dominated by the outright forward because transactions costs are doubled. Specifically, we consider option contracts with maturities ranging from one to 24 months. So, at maturity, an option which has been purchased at some decision stage is either exercised, if it yields a positive payoff, or is simply left to expire. Option strike prices for European style straddle options with maturities from 1 month to 24 months, as depicted in Fig. 15.6, were specified exogenously as inputs to the model and were obtained from the Bloomberg database [44].

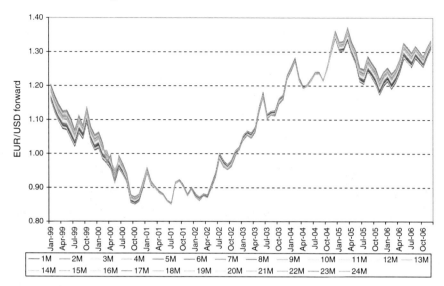

Fig. 15.5 At-the-money forward rates

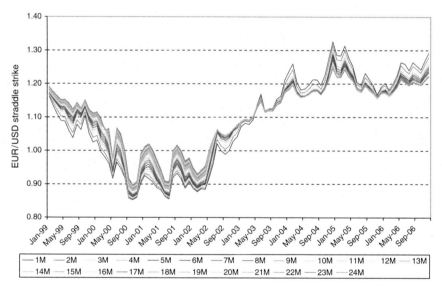

Fig. 15.6 At-the-money strike prices for European style straddle options

Strike prices for given maturities were chosen by taking the arithmetic mean between the firm's expected exchange rate for that maturity as given by the PPP reverting simulation model, and the exchange rate as expected by the market which is given by the at-the-money straddle premium. It is only under this circumstance that the firm and the capital market expect equal payoffs and hence, the straddle can be considered as fairly valued from the firm's point of view [323].

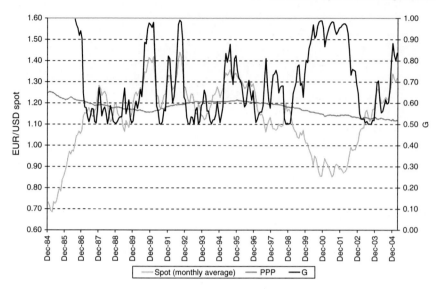

Fig. 15.7 Historical transition function estimates

15.2.2.4 Simulation Parameters

An exchange rate scenario is a sequence of values for a single exchange rate over the analysis period. The purpose of scenario generation is to come up with levels for the exchange rate path between $t = 0$ and $T = 24$. For a single factor the discrete form of the evolution equation for the exchange rate s_t is given by (14.7). For ease of model calibration, we assume $c_1 = c_2 = PPP$. The variables γ and σ were estimated on the 15th of every month according to (14.8) and (14.9) for the time period between January 1985 and December 2004 (252 months).

Figure 15.7 illustrates the resulting estimated values for the transition function G over the backtesting period. Obviously, G is strongly reacting to the degree of disequilibrium. Figure 15.8 displays the estimated values for σ, where each estimate refers to the standard deviation of EUR/USD moves of the past 24 months. Since only one exchange rate is considered there is no need to make assumptions about dependence structures.

15.2.2.5 Optimization Parameter Settings

In order to derive decisions on the mix of instruments, we consider the SCOP problem with a mean-variance-skewness objective as formulated in Problem 4.1. The decision variables are represented by the amount of the hedge that is to be allocated to each of the hedging vehicles. The optimal solution is represented by the allocation of instruments $x^* = (x_F^*, x_O^*, x_S^*)$ that yields the minimal objective function value. For exposition purposes, the optimal weight sets used to construct the hedges given

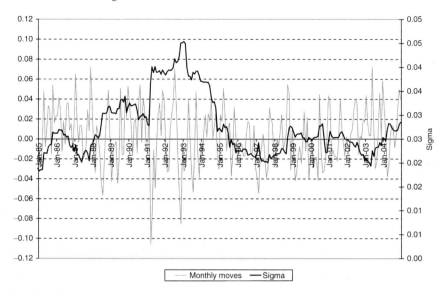

Fig. 15.8 Historical standard deviation estimates

the firm's preference are set to $\lambda = \kappa = 1$. This preference set is a compromise case where the weights for mean, variance, and skewness are equal, indicating that mean, variance, and skewness of return are of equal importance to the firm. The starting values for the decision variables were set to $x_F^* = 0.33$, $x_O^* = 0.33$, and $x_S^* = 0.34$ in order to provide a neutral start for the search. The step size for the decision variables was set to 0.01, and according to Problem 13.1, 0 and 1 provide the lower and upper bounds for x_F, x_O, x_S whose sum must equal 1. We further specify that the search stops after 150 simulations with each simulation comprising 10,000 replications of the exchange rate with the same initial population given a specific pseudorandom generator seed. The latter reduces the variance between runs and makes the search for an optimum more efficient. All tests were carried out on a standard desktop PC (2 GHz single CPU) with Windows XP.

15.2.2.6 Algorithmic Reference

The commercial Monte Carlo spreadsheet add-in, Crystal Ball version 7.02, was used for generating exchange rate scenarios.[1] To estimate optimal values for the decision variables, we applied the scatter search/path relinking algorithm according to [238]. The algorithm is implemented in the commercial OptQuest package, a stand-alone optimization routine that is bundled with Crystal Ball. Being a completely separate software package, OptQuest treats the simulation model as a black box, i.e., it observes only the input/output of the simulation model.

---

[1] Crystal Ball's random number generator has a period of length $2^{31} - 2$ which means that the cycle of random numbers repeats after several billion trials.

15.2.3 *Evaluation Procedure*

The role of the historical market data, the exposure data, their interaction with the decision model and backtesting are described in the following. The experimental setup requires that we dynamically:

1. Use market data in order to revalue the forward and straddle positions
2. Record the decisions made in the current step of the model as an input of the starting position of the next "roll" of the model

As we stepped through time the model database was updated with the most currently available equilibrium, spot, forward and straddle rates. Once the current forward exposures for the next 24 months were accessed, the favorability of the current exchange rate was evaluated resulting in a recommendation on the future exposure to be hedged, i.e., a target hedge intensity. If this number was equal or higher than the current amount of months hedged, a decision on the composition of the hedge had to be made. In this case, OptQuest was used to solve the resulting optimization problem. For instance, consider the situation on 15 November 2000. The suggested hedge intensity according to Fig. 15.4 is 19 months. Since 18 months have been hedged in the past, a decision on the instrument mix had to be made for month 19, i.e., June 2002 only.

Figure 15.9 shows the performance graph that resulted from an OptQuest run of 100 simulations. The graph shows the value of the objective for the best alternative found as the search progresses. The search began from a solution with an expected objective function value of 2.5454, which happened to be a goal-feasible solution. The best solution was found on the 70th simulation and had an objective function value of 2.5522. One can also observe that the best value changes quickly early

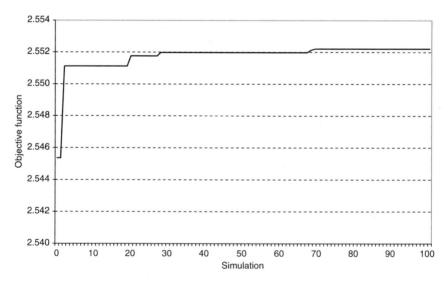

Fig. 15.9 Performance graph

Table 15.1 Best solution found

Simulation	Objective	Spot	Forward	Straddle
1	2.54536	0.34	0.33	0.33
2	2.55112	0.00	0.50	0.50
20	2.55176	0.00	0.46	0.54
28	2.55198	0.00	0.44	0.56
68	2.55213	0.00	0.42	0.58
69	2.55221	0.00	0.40	0.60
70	2.55222	0.00	0.38	0.62

Table 15.2 Robustness of best solution found

Name	Best	Minimum	Average	Maximum	Std. Dev.
Objective	2.5522	2.5204	2.5414	2.5522	0.0093
Spot	0.00	0.00	0.2336	1.00	0.2954
Forward	0.38	0.00	0.3526	1.00	0.2681
Straddle	0.62	0.00	0.4139	1.00	0.2904

in the run, as improvements are being found frequently, and then it changes more slowly as better solutions become harder to find. We find in our experiments that scatter search is a search method that is both aggressive and robust. It is aggressive because it finds high-quality solutions early in the search. It is robust because it continues to improve upon the best solution when allowed to search longer.

The details of this run are reported in Table 15.1, where each row gives the values of the decision variables in case they led to a higher objective function value. The best solution was achieved in the 70th run which recommends to hedge the exposure completely with a combination of 38% forwards and 62% straddles. These selections meet the category bounds specified in Problem 13.1, and provide a single good solution, thereby guiding policy for increasing mean-variance-skewness.

Given a set of solutions, all possible of increasing mean-variance-skewness, it is interesting to ask how similar these solutions are in terms of their suggested allocations. If all these solutions suggest a similar strategy, then one can be more confident that this policy is robust to small changes in the amounts allocated. Alternatively, if these solutions correspond to extremely different allocations, then this implies that the solution is extremely sensitive to the amounts allocated. We identified 101 solutions that had an objective within 10% of the best objective.

The solution report given by Table 15.2 provides evidence that the objective function value is relatively noisy. This is not the result of a low number of simulation trials but due to the relatively wide spreads between the minimum and maximum values for the decision variables from the set of solutions that fell within the analysis range.

Furthermore, we analyze the runtime behavior of the respective algorithm by the required computing time on the reference PC. Under the settings described, solving the problem for three decision variables took approximately 12 min and solving the problem for six decision variables came at the expense of approximately 20 min.

On the one hand, the computing time should not be overinterpreted because it is dependent on the reference machine and on implementation details, but on the other hand, it is an indication for real-world applications.

In order to validate the model over time, we calculated its effect on net income via a rolling window approach where the objective function is given by

$$U(w) = \sum_{t=1}^{T} U(w_t) \tag{15.1}$$

As per 31 December 1998, we assumed that the firm's net income was USD 100. As per 15 January 1999, a hedge intensity of ten months (covering February 1999 to November 1999) was suggested by the procedure described in Fig. 15.4. Nine months, i.e., February 1999 to October 1999, were assumed to be covered in the past, each by a proportion of 1/3 forwards and 1/3 straddle options. The remaining 1/3 of the following 9 months' exposure, along with the exposures from month 11 to month 24, were assumed to remain unhedged. In order to obtain a recommendation on the composition of the hedge for maturity November 1999, the simulation/optimization procedure was started. Upon determination of the search, the best values for the decision variables as obtained by OptQuest were recorded. The cost of the hedge was calculated (straddle premium + transaction cost of 3 pips) and deducted from last month's net income. Next, the clock was advanced by one month, and the financial losses or gains made on the forward and straddle positions were determined. Thus, on 15th January 1999, the first day of the backtesting period, all cashflow maturities have decreased by one month (i.e., the exposure in month 2 becomes an exposure in month 1, the exposure in month 3 becomes an exposure in month 2, etc.) and a new exposure in month 24 appears in the 24 months time window. In realigning the forward and straddle contracts to the current spot levels either some profit on the currently held contracts is realized or the selective hedge has led to a loss. Since the decision model uses data sets, which were updated as we stepped through time, the scenario generator created a completely new set of scenarios in each month looking ahead over a time horizon of $T = 24$ months. Given that the decisions were made altogether 72 times, by stepping through the time line we processed the corresponding simulation/optimization model 72 times using the Crystal Ball/Optquest system. To verify whether multiproduct hedging with variable weights has advantages over simpler approaches, the dynamic rolling procedure was not only applied to the mean-variance-skewness optimization model but also to four benchmark strategies:

1. 100% spot strategy (no hedging), i.e., $x_S = 1$ and $x_F = x_O = 0$, for $t = 0, \ldots, T$
2. 100% forward strategy, i.e.,

- $x_F(t^*) = \begin{cases} 1 & : \quad t = 0, \ldots, t^* \\ 0 & : \quad t = t^*, \ldots, T \end{cases}$
- $x_O = 0$ for $t = 0, \ldots, T$
- $x_S(t^*) = \begin{cases} 0 & : \quad t = 0, \ldots, t^* \\ 1 & : \quad t = t^*, \ldots, T \end{cases}$

3. 100% straddle strategy, i.e.,

- $x_F = 0$ for $t = 0, \ldots, T$
- $x_O(t^*) = \begin{cases} 1 & : & t = 0, \ldots, t^* \\ 0 & : & t = t^*, \ldots, T \end{cases}$
- $x_S(t^*) = \begin{cases} 0 & : & t = 0, \ldots, t^* \\ 1 & : & t = t^*, \ldots, T \end{cases}$

4. Mixed strategy consisting of 1/3 at-the-money forwards, 1/3 at-the-money straddle call options, and 1/3 spot, i.e.,

- $x_F(t^*) = \begin{cases} 0.33 & : & t = 0, \ldots, t^* \\ 0 & : & t = t^*, \ldots, T \end{cases}$
- $x_O(t^*) = \begin{cases} 0.33 & : & t = 0, \ldots, t^* \\ 0 & : & t = t^*, \ldots, T \end{cases}$
- $x_S(t^*) = \begin{cases} 0.34 & : & t = 0, \ldots, t^* \\ 1 & : & t = t^*, \ldots, T \end{cases}$

Strategies 1–4 are nonpredictive beyond the given grid recommendation since they do not allow the hedge to time-vary but are fixed-weight strategies and therefore do not seek to exploit the history of market information.

15.3 Results

15.3.1 Ex Ante Performance

Figure 15.10 plots the objective function values of the proposed mean-variance-skewness utility function against the four benchmark strategies over the 8 year out-of-sample backtesting period from January 1999 to December 2006. Grey bars indicate times when decisions on the optimal combination of instruments were delayed in order to reduce hedge intensities.

On 15 December 1998 – as mentioned above – all four strategies start with same initial product mix, a hedge intensity of ten future months (covering February 1999 to November 1999) with each of them being covered by a proportion of 1/3 forwards and 1/3 straddle options. The remaining 1/3 of the following 9 months' exposure along with the exposures from M11 to M24 remained unhedged.

We observe that from here on the utility of the spot strategy gradually declines over the next 9 months, and in September 1999 jumps down to a utility value below zero. Since no further hedges are made, the strategy has the highest variance consistently over time which is punished by the objective function. In addition, since it does not represent a hedge, the payoff distribution from which the mean and skewness values are derived becomes irrelevant. This explains the smooth path of the spot strategy at a negative level of utility.

Visual inspection of Fig. 15.10 further reveals that the forward strategy provides the most volatile objective function values taking on relatively high – albeit

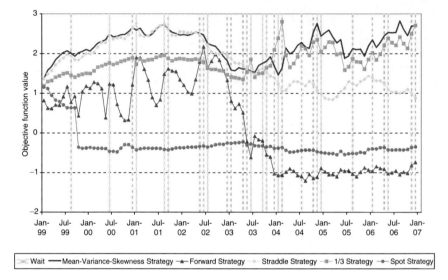

Fig. 15.10 Historical comparison of objective function values

volatile – utility values from January 1999 to May 2003, and a low level of utility from May 2003 onwards. This can be explained as follows. According to PPP, the USD is strong compared to the EUR from January 1999 to May 2003. Relatively few USD have to be given up in order to buy one EUR. Since the USD is overvalued, we expect that it will become weaker in the future which is expressed by the stochastic smooth transition PPP process incorporating a nonlinear positive trend in the exchange rate scenarios towards equilibrium. This in turn implies that forward hedges implemented in the past are *in-the-money* which results in a positive mean of the payoff-distribution. From May 2003 onwards, the exact opposite holds. The USD overvaluation cycle has ended and the spot rate enters into a phase of USD undervaluation. The mean of the probability distribution of payoffs becomes negative because it is expected that – due to gravitation – future spot is likely to go down at a pace γ that is linked to the degree of USD undervaluation and forward hedges implemented in the past are therefore *out-of-the-money*. To sum up, entering into a forward contract always includes a directional bet on the future path of the spot exchange rate while the variance of the open position is symmetrically reduced and third-moment effects do not exist.

The straddle strategy clearly dominates both the spot and forward strategies throughout the whole period. The reason is twofold. First, like forwards, straddles eliminate exchange rate risk as measured by σ. Second, unlike forwards, they provide a profit/loss potential beyond the forward contract depending on whether exchange rate volatility expectations were closer to reality ex post than those of the counterparty selling the straddle. If exchange rate volatility was expected to be higher than implied volatility in the capital market, this had a direct positive impact on the mean of the straddle contract's payoff distribution, and vice versa. From

January 1999 to May 2003, spot upward movements were predicted to be large enough by the smooth transition PPP model to activate the call option of the straddle and offset the premium paid. On average, the contribution of the mean payoff on the objective function value was higher than the skewness contribution. From May 2003 onwards, EUR/USD downward movements were predicted to be large enough to activate the put option of the straddle which kept the strategy profitable. Still, expected straddle profits were not expected to be as high as the ones in the first half of the backtesting period. One reason is that the estimated values for γ continually decline within our estimation procedure as a result of averaging. This results in a slowing adjustment and therefore implies smaller profit potential for straddles. The other reason is that hedge intensities were reduced during this period which implies smaller profit potential due to less hedges. Instead, the skewness contribution increased in the second half.

A comparison of the strategies confirms our expectation that the variable-weight mean-variance-skewness strategy's utility level (black line) dominates the traditional fixed weight allocation prescriptions over time. The optimized mean-variance-skewness strategy is closely mirrored by the straddle strategy. Then, from August 2003 onwards the two strategies diverge and the proposed strategy is better mirrored by the 1/3 strategy. This is a natural result of the scatter search optimization routine tending to select the best of both worlds: completely closed positions up to the recommended hedge intensity with an emphasis on straddles during the USD overvaluation period from January 1999 until April 2003, and just as many hedges as necessary to suffice the given degree of risk-aversion during the USD undervaluation. In particular, it was the skewness criterion that drove decisions from April 2003 – the turning point in the backtesting period when the USD overvaluation (EUR undervaluation) cycle turns into a USD undervaluation (EUR undervaluation) phase – onwards. Although in the aftermath of April 2003, the mean of the straddle's payoff distribution is consistently higher than the mean of the optimized payoff distribution, this comes at a tremendous cost of skewness. This seemed at first surprising since the optimized strategy picks – on average – less straddle contracts (-21%), but a higher proportion of forwards ($+10\%$) and a noticeably higher spot proportion ($+11\%$) instead. Hence, despite the strategy containing less options, it is more positively skewed since the probabilistic losses of the forward proportion offset part of the straddle's probabilistic profits (asymmetric distribution). Further note, that in the first half of the backtesting period there were months when the optimized strategy fell below the straddle strategy, and in the second half of the backtesting period there exist months when the optimized strategy fell below the mixed strategy of 1/3 spot, 1/3 forwards, and 1/3 options. The reason is, that at times when hedging intensities were reduced ("do nothing"), i.e., exposures were left uncovered and completely exposed to spot rate fluctuations, worse objective function values were achieved due to lower skewness if the optimized strategy included less straddle contracts than the competitive strategies. Overall, the straddle, forward, and spot strategies perform significantly worse than the mixed strategies. The problem with the straddle strategy is that option premia are expensive and reduce the mean, in particular, at times when volatility in the market was lower than expected. In contrast, the forward's purpose to cheaply reduce or eliminate income volatility turns out to be problematic if spot moves in the opposite direction. One can benefit from such moves by following a pure spot strategy and leave the exposure totally open, however the high variability of results is punished by the objective function.

15.3.2 Ex Post Performance

An ex post comparison of the different hedging strategies' net income is given in Fig. 15.11.

There is no dominant strategy that outperforms the other ones constantly over time. However, both the straddle strategy and the optimized mean-variance-skewness strategy seem to be clearly preferable over the other strategies. All five strategies started with a net income of USD 100 which was reduced to USD 99.82 in January 1999 as a result of paying straddle option premia on 1/3 of the exposed amounts for the next 9 months. The simplest approach, presented by the red line is that of no hedge at all. In this case, the portfolio simply comprises a long position in the spot market having zero payoff from October 1999 on, when the initial positions were settled.

The forward strategy leads to the most volatile P&L as a result of the forward bias, a phenomenon vastly described in both academic [114, 125] and practical literature [29]. What this literature implies is that historically, rather than the USD depreciating as suggested by the forward rate, the USD was actually appreciating from January 1999 until the beginning of 2002 causing large losses of the forward

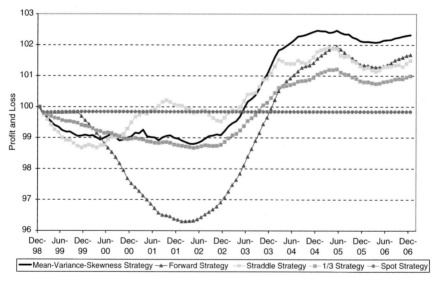

Fig. 15.11 Historical comparison of profit and loss accounts

strategy at expiry. This led to a minimum net income of USD 96.3 in January 2002. In contrast, positive returns were achieved by the forward strategy over the following months due to a forward premium discount bias, i.e., the USD turned out to be weaker as initially expected by the forward rate. Hence, the realized net income recovered since January 2002 exceeding the initial USD 100 two years later in January 2004, reaching the maximum of USD 101.95 in May 2005, and moving sideways until the end of the backtesting period. The 1/3 strategy is the least volatile in terms of profit/loss, mainly due to its large spot proportion which limits upside, as well as downside profit potential. The straddle strategy seems to provide very desirable results over time. One possible explanation is that the capital markets underestimated future EUR/USD volatility when pricing the options. Santa-Clara and Saretto [354] examined various strategies, including naked and covered position in options, straddles, strangles, and calendar spreads, and find support for mispricing in the options markets which cannot be arbitraged away due to high trading costs and margin requirements.

The mean-variance-skewness strategy generates a growth path with similar patterns to those of the 1/3 strategy until December 2002, then rises strongly and outperforms the 1/3 strategy and even the straddle strategy. The latter fact comes at the expense of increasing volatility. Therefore, the firm implementing the multiproduct optimal hedging model will be significantly less than fully hedged. One possible interpretation of the better performance of the dynamically optimized strategy over the naive hedges is that it uses short-run information, while the naive static hedges are driven by long-run considerations and an assumption that the relationship between spot and forward price movements is 1:1. For instance, if the indication is in favor of a USD appreciation, then being less than fully hedged will lead to gains in the spot market conversion, and lower losses (or no losses) on the forward contract leading to much better financial results. On the other hand, if the indication is in favor of a depreciation of the foreign currency, then being hedged is preferable to being unhedged and will lead to gains in the forward contracts.

An alternative way of presenting ex post results of the rolling decision model is given by Fig. 15.12 which provides a histogram of the different strategies' realized profit/loss. In addition to Fig. 15.12, Table 15.3 provides the associated summary statistics of the backtested hedging strategies' probability distributions.

The forward strategy has the longest left tail and implies the highest risk of a shortfall in net income. With a standard deviation of 1.95, it is also the most volatile strategy. From a risk-return perspective, the forward strategy has the lowest Sharpe Ratio and the weakest return per unit of earnings-at-risk (EaR) which was calculated as the difference between the initial amount USD 100 and the 5% quantile of the respective probability distribution of realized profit/loss. The straddle strategy and the 1/3 strategy reveal similar risk-return characteristics if compared by their Mean/EaR ratios with the straddle strategy at 79.84 being slightly better. However, if compared against traditional Sharpe Ratio, superior results are provided by the 1/3 strategy which is due to the lowest standard standard deviation among all strategies. The optimized mean-variance-skewness strategy provides superior out-of-sample results if profit/loss risk is not perceived symmetrically but in terms of downside risk. In fact,

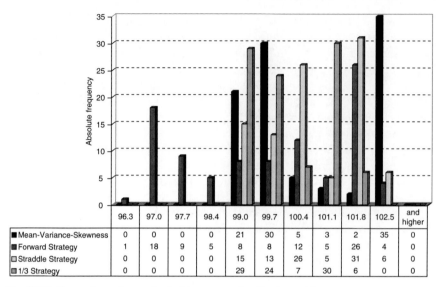

Fig. 15.12 Comparison of strategies by histograms of realized monthly revenues

Table 15.3 Summary statistics of the hedging strategies backtested

	Mean	SD	EaR	Min	Max	Mean/SD	Mean/EaR
M-V-S	100.40	1.49	1.10	98.80	102.47	67.45	91.28
Forward	99.29	1.95	3.64	96.30	101.95	50.93	27.28
Straddle	100.32	1.02	1.26	98.68	101.94	97.88	79.84
1/3	99.82	0.90	1.27	98.67	101.21	110.95	78.75

most corporations and investors would be foolish to not accept the good risk, but would be negligent to not eliminate the bad risk.

Still, the explanatory power of the results is limited, especially if we consider that exchange rate adjustment is a long-term phenomenon. We can only get a vague idea on whether the forecasts from a STAR model are always superior to those of the linear models out-of-sample. Evaluating forecast performance is therefore intrinsically problematic. As noted by [403], even though nonlinear time series often capture certain characteristics of the data better than do linear or random walk models, the forecast performance of the former is not always superior, and is sometimes even worse. Hence, there is no guarantee that the estimated LSTAR models will produce superior forecasts. A necessary condition for that to happen would seem to be that the forecasting period contains *nonlinear features* [180]. For instance, a nonlinear model may be expected to be superior to a linear one when the forecasting period contains the aftermath of a large negative shock. If that is not the case, a linear autoregressive model is likely to perform as well as a nonlinear one. Another possible reconciliation is offered by [89], who suggest that regime switching models may forecast poorly owing to the difficulty of forecasting the regime that the series will be in. Thus, any gain from a good fit of the model within the regime will be lost if the model forecasts the regime incorrectly. Estimated values of parameters μ, σ, and γ in the exchange rate process using MCS are therefore indicative but not perfectly predictive of similar parameters for future exchange rate returns. This will be true for the parameters of any parametric exchange rate evolution model. Therefore, there is an inherent uncertainty associated with mean, standard deviation, adjustment speed, etc. of future exchange rate returns. As a result, the apparent precision of an exact solution is somewhat deceptive since the problem may itself be viewed as not exactly defined due to lack of precision in the problem parameters. Therefore, precision of the parametric solution is a problem only if the parameters for exchange rate dynamics are known with greater precision. In case parameters are known with higher precision, a different parameterization may be used for a better approximation. This will however, result in a larger problem size. Despite the many problems that economic forecasts from economic systems confront, these models offer a vehicle for understanding and learning from failures, as well as consolidating our growing knowledge of economic behavior.

Part V
Contributions of the Dissertation

Chapter 16
Exchange Rate Forecasting with Support Vector Machines

This thesis is an interdisciplinary work that combines computer science, econometrics and financial engineering in order to support human decision-making in the context of trading and hedging the foreign exchange market. The first part of the dissertation provides an introduction to basic macroeconomic and financial theories on exchange rates. While this part does not advance any new ideas, it presents an original synthesis of existing work from different research directions, stressing the relevance of computational complexity theory to address financial market phenomena which cannot be explained by conventional financial and macroeconomic theories. The remainder of the dissertation makes the following contributions.

Abstract – Purpose: Part II examines and analyzes the general ability of support vector machine (SVM) models to correctly predict and trade daily EUR/GBP, EUR/JPY, and EUR/USD exchange rate return directions. When computers are applied to solve a practical problem, it is usually the case that the method of deriving the required output from a set of inputs can be described explicitly. As computers are applied to solve more complex problems, however, situations can arise in which there is no known method for computing the desired output from a set of inputs, or where that computation may be very expensive. Forecasting financial time series events such as daily exchange rate directions, for instance, is a problem that is very relevant for the financial community and known to be very difficult in practice. We formally represent this problem as a classification task which is described by the linear separability problem. In the special case of finding whether two sets of points (namely exchange rate ups and downs) in general space can be separated, the linear separability problem becomes the binary classification problem whose most general form, the case of whether two sets of points in general space can be separated by k hyperplanes, is known to be NP-complete. It is generally believed that NP-complete problems cannot be solved efficiently.

Design/Methodology/Approach: We approach the task of solving the BCP in the context of predicting daily exchange rate directions with supervised learning, a learning methodology for the computer that attempts to learn the input/output functionality from historical training examples. In particular, we opt for SVM as a supervised learning algorithm. SVM has been found to work well for classification tasks across a variety of scientific disciplines. The basic idea is to project the input data via kernel into a more expressive, high dimensional feature space where the SVM algorithm finds a linear decision plane that has maximum distance from the nearest training patterns. Using a kernel function guarantees that linear classification in feature space is equal to nonlinear classification in input space. Our experiment is to compare six SVM models with varying standard kernels along with one exotic p-Gaussian SVM in order to investigate their quality in separating Granger-caused input data in high dimensional feature space. To ascertain their potential value as out-of-sample forecasting and quantitative trading tools, all SVM models are benchmarked against traditional forecasting techniques.

Practical Implications: Apart from pure statistical evaluation, the performance of SVM was tested in a real-world environment. We set up a trading simulation where return predictions were translated into positions first. Next, a decision framework was established that indicates when the underlying asset was bought or sold, depending on the SVM output. Several trading metrics were imposed on the forecasting results to measure the models' success.

Results/Findings: It is found that hyperbolic SVMs perform consistently well in terms of forecasting accuracy and trading performance via a simulated strategy. Moreover, p-Gaussian SVMs perform reasonably well in predicting EUR/GBP and EUR/USD return directions.

Originality/Value: The results of Part III shed light on the existence of a particular kernel function which is able to represent properties of exchange rate returns, generally well, in high dimensional space. The fact that hyperbolic kernels are such promising candidates can be valuable for institutional investors, private investors, and risk managers of nonfinancial corporations.

Future research: Future research direction will focus on SVM models and their technical improvements. In light of this research, it would also be interesting to see if the dominance of hyperbolic SVMs can be confirmed in further empirical investigations on financial market return prediction. We believe that modern methods of the machine learning, pattern recognition, and empirical inference will play an essential role in coping with a complex world.

Chapter 17
Exchange Rate Hedging in a Simulation/Optimization Framework

Abstract – Purpose: Part III considers another relevant practical problem involving uncertainty about future exchange rate developments: the problem of finding a possibly optimal combination of linear and nonlinear financial instruments in order to hedge foreign exchange transaction risk over a specified planning period. While much research has been done in order to find answers on why multinational firms in reality hedge their exchange rate exposures, newspapers regularly suggest that more research should be directed towards how hedgers can improve their decision making procedures. Hedges are trades designed with the motivation to reduce or eliminate risk which distinguishes them from pure speculation. Still real-world hedging policies often involve a speculative component in terms of exchange rate expectations which may deviate from those implied by derivative prices. For instance, if a hedger believes that the currency is going to move in an unfavorable direction, he may ask about the appropriate strategy in terms of instrument selection. In contrast, if the currency is expected to move favorably, but the hedger is not entirely sure, should he use a different risk management strategy?

Design/Methodology/Approach: We address the conflicting empirical finding that firms do like to try to anticipate events, but that they also cannot base risk management on second-guessing the market. Our analysis therefore argues that a way to understand corporate hedging behavior is in the context of speculative motives that could arise from either overconfidence or informational asymmetries. The problem is addressed in a formal way, taking into account one single source of uncertainty (the exchange rate), a given set of instruments consisting of spot, forward, and European straddle option contracts that the hedger is allowed to choose from, and his particular attitudes towards risk and return. Regarding the latter, financial literature has shown that traditional mean-variance or mean-quantile-based performance measures may be misleading if products with nonlinear payoff profiles, such as options, are used. In addition, literature on bounded rationality suggests that individuals perceive risk in a nonlinear fashion which is expressed by a preference for positive skewness in the probability distribution of future payoffs. For these reasons, a weighted mean-variance-skewness utility maximization framework with linear

C. Ullrich, *Forecasting and Hedging in the Foreign Exchange Markets,* Lecture Notes in Economics and Mathematical Systems 623, DOI: 10.1007/978-3-642-00495-7_17, © Springer-Verlag Berlin Heidelberg 2009

constraints is embedded in a single-period stochastic combinatorial optimization problem (SCOP) formulation with the goal to maximize expected utility at the planning horizon. Stochastic combinatorial optimization is the process of finding the best, or optimal solution for problems with a discrete set of feasible solutions in a stochastic system. The single-period approach suggests that hedge ratios cannot be altered once hedging decisions are made. While much progress has been made in finding exact (proveably optimal) solutions to some combinatorial optimization problems, using techniques such as dynamic programing, cutting planes, and branch-and-cut methods, reaching "optimal solutions" is in many cases meaningless, as in practice we are often dealing with models that are rough simplifications of reality. The underlying SCOP is proveably *NP*-hard and therefore, too difficult to be solved analytically in polynomial time which requires the use of good heuristic methods. In addition, the optimization problem exhibits multiple local extrema and discontinuities. In the absence of a tractable mathematical structure, we study the behavior of heuristics, "quick and dirty" algorithms, which return feasible solutions that are not necessarily optimal. In particular, we apply a methodology called Simulation/Optimization which is a general expression for solving problems where one has to search for the settings of controllable decision variables that yield the maximum or minimum expected performance of a stochastic system as presented by a simulation model. For modeling the EUR/USD exchange rate, a smooth transition nonlinear PPP reversion model is presented. The simulation model is very attractive in the present context and unique to our knowledge. It addresses both, the first and the second PPP puzzle, and provides a theoretically valid and visually intuitive view on the corridor of future EUR/USD spot development. The key feature is a smooth transition function which allows for smooth transition between exchange rate regimes, symmetric adjustment of the exchange rate for deviations above and below equilibrium, and the potential inclusion of a neutral corridor where the exchange rate does not mean revert but moves sideways. Another advantage is that only two parameters need to be estimated, the speed of mean reversion and exchange rate volatility, which makes the model straightforward to use in practice. For the task of optimization, we propose the use of a metaheuristic combinatorial search algorithm. A metaheuristic refers to an intelligent master strategy that guides and modifies other heuristic methods to produce solutions beyond those that are normally generated in a quest for local optimality. The specific metaheuristic we use is a variant of scatter search, a generalized form of path relinking, which in many practical problems has proven to be an effective and efficient approach due to its flexibility to accommodate variations in problem structure and in the objectives considered for the evaluation of solutions.

Practical Implications: In order to show our simulation/optimization model's applicability in a practical context, a situation is presented, where a manufacturing company, located in the EU, sells its goods via a US-based subsidiary to the end-customer in the US. Since it is not clear what the EUR/USD spot exchange rate will be on future transaction dates, the subsidiary is exposed to foreign exchange transaction risk under the assumption that exposures are deterministic. We take the view that it is important to establish whether optimal risk management procedures offer a significant improvement over more ad hoc procedures. We demonstrate that the establishment of simple rules which adjust the recommended hedge dynamically (without rebalancing) based on easily observed current market factors can lead to better performance and risk management than various nave static hedges. For the purpose of model validation, historical data backtesting was carried out and it was assessed whether the optimized mean-variance-skewness approach is able to outperform passive strategies such as unitary spot, forward, and straddle, as well as a mixed strategy over time. The passive strategies are nonpredictive since they do not allow the hedge to time-vary but are fixed-weight strategies and therefore do not use the history of market information. We compare the alternative strategies in dynamic backtesting simulations using market data on a rolling horizon basis. The strategies were evaluated both in terms of their ex ante objective function values – as well as in terms of ex post development of net income.

Findings: Financial markets are perfect examples of dynamic systems of complex human behavior. Unsurprisingly to either algorithm designers or theorists, we provide proveable results that incomplete information about the state of the system hinders computation. Although we model only one "simple" random price movement, the EUR/USD exchange rate, our results demonstrate that currrency hedging in practice may be a hard problem from a computational complexity perspective and that optimal solutions to real-world hedging problems can be approximated at best. We demonstrate through extensive numerical tests the viability of a simulation/optimization model as a decision support tool for foreign exchange management. We find in our experiments that scatter search is a search method that is both aggressive and robust. It is aggressive because it finds high-quality solutions early in the search. It is robust because it continues to improve upon the best solution when allowed to search longer. Our approach to hedging foreign exchange transaction risk is based on exchange rate expectations, considers real market data and incorporates flexible weights. We find that the approach adds value in terms of reducing risk and enhancing income. The optimized mean-variance-skewness strategy provides superior risk-return results in comparison to the passive strategies if earnings risk is perceived asymmetrically in terms of downside risk. Conditioning information therefore seems to be important. Even with low levels of predictability, there is a substantial loss in opportunity when fixed-weight strategies (which assume no predictability) are implemented relative to the dynamic strategy that incorporates conditioning information. The pure forward strategy is found to have the lowest return per unit of earnings risk whereas the straddle strategy and the 1/3 strategy reveal similar risk-return characteristics. Interestingly, our research also contrasts the finding that currency forward contracts generally yield better results in comparison to options since a passive straddle strategy would have yielded superior results compared to a forward strategy. Our results are as follows. Apart from our backtesting results, it is believed that the proposed simulation/optimization procedure for determining optimal solutions has important implications for policy making. Having easy access to relevant solutions makes it easier for policy makers to explore and experiment with the model while incorporating their own intuition before deciding on a final plan of action. Despite the many problems that economic forecasts from economic systems confront, these models offer a vehicle for understanding and learning from failures, as well as consolidating our growing knowledge of economic behavior.

Future research: The only way to guarantee the best possible combination is to know in advance whether the currency will move in one's favor or not. Hence, there will always be a natural desire to foresee the future by improving on forecasting. This may require more complex models than our "simple" stochastic representation of the exchange rate. Where our model only describes statistically some restricted aspects of this rich complexity, real-world complexity is associated with a range of behaviors and functions that go way beyond our approximation. Thus, it can be argued that more complex behavior needs more complex models. For instance, one important additional element adding to real-world complexity is that risk management objectives are often perceived towards different reference levels, such as exposure, target exchange rates as established from their own budgeting and planning processes, or even personal anchor points that may involve expert forecasts. Second, hedging policies are often benchmark-orientated, position adjustment is done in terms of market values, and exposures are uncertain. Third, a hedger's perception of risk may depend on different states of nature, which makes it a multidimensional phenomenon. These considerations call for a more in-depth model of hedging, which would allow the hedge ratios to have a stochastic nature. Another, more popular field of research, involves modeling departures from the normal distribution of exchange rate returns in order to better capture the probability of tail events. Throughout Part IV, we assumed that the individual utility function is known and given as an easily applicable formula. We therefore ignored the difficulty of learning one's utility function which may consitute another interesting real-world problem.

Part VI
References

References

1. Abeysekera S.P., Turtle H.J.: Long-Run Relations in Exchange Markets: A Test of Covered Interest Parity. J. Financial Res. **18(4)**, 431-447 (1995)
2. Abuaf N., Jorion P.: Purchasing Power Parity in the Long Run. J. Finance **45**, 157-174 (1990)
3. Acerbi C., Tasche D.: On the Coherence of Expected Shortfall. J. Bank. Finance **26(7)**, 1487-1503 (2002)
4. Acerbi C., Nordio C., Sirtori C.: Expected Shortfall as a Tool for Financial Risk Management. Working Paper (2002) http://www.gloriamundi.org/var/wps.html. Cited 12 June 2005
5. Adler M., Dumas B.: International Portfolio Choice and Corporation Finance: A Synthesis. J. Finance **37(3)**, 925-984 (1983)
6. Adler M., Lehmann B.: Deviations from Purchasing Power Parity in the Long Run. J. Finance **38(5)**, 1471-1487 (1983)
7. Albuquerque R.: Optimal Currency Hedging. EconWPA Finance 0405010 (2004) http://ideas.repec.org/p/wpa/wuwpfi/0405010.html. Cited 20 Nov 2005
8. Anderson J.R.: Is Human Cognition Adaptive? Behav. Brain Sci. **14**, 471-484 (1991)
9. Andreou A.S., Georgopoulos E.F., Likothanassis S.D.: Exchange-Rates Forecasting: A Hybrid Algorithm Based on Genetically Optimized Adaptive Neural Networks. Comput. Econ. **20(3)**, 191-210 (2002)
10. Arditti, F.D.: Risk and the Required Return on Equity. J. Finance **22(1)**, 19-36 (1967)
11. Aronszajn N.: Theory of Reproducing Kernels. Trans. Am. Math. Soc. **68**, 337-404 (1950)
12. Arrow K.J.: Aspects of the Theory of Risk Bearing. Yrjo Jahnsson Lectures, The Academic Book Store, Helsinki (1965)
13. Arrow K.J., Debreu G.: Existence of an Equilibrium for a Competitive Economy. Econometrica: J. Econ. Soc. **22(3)**, 265-290 (1954)
14. Arthur W.B.: Inductive Behaviour and Bounded Rationality. Am. Econ. Rev. **84**, 406-411 (1994)
15. Arthur W.B.: Complexity and the Economy. Sci. **284**, 107-109 (1999)
16. Arthur W.B., Holland J., LeBaron B., Palmer R., Taylor P.: Asset Pricing Under Endogenous Expectations in an Artifical Stock Market. In: Arthur W.B., Durlauf S., Lane D. (eds.) The Economy as an Evolving Complex System II, pp. 15-44. Addison-Wesley, Reading MA (1997)
17. Artzner P., Delbaen F., Eber J.-M., Heath D.: Coherent Measures of Risk. Math. Finance **9**, 203-228 (1999)
18. Atan T., Secomandi N.: A Rollout-Based Application of the Scatter Search/Path Relinking Template to the Multi-Vehicle Routing Problem with Stochastic Demands and Restocking. PROS Revenue Management, Houston TX (1999)
19. Axtell R.: The Complexity of Exchange. Econ. J. **115(504)**, 193-210 (2005)
20. Baillie R.T., Bollerslev T.: The Forward Premium Anomaly is not as bad as you think. J. Int. Money Finance (19), 471-488 (2000)

21. Baillie R., McMahon P.: The Foreign Exchange Market: Theory and Econometric Evidence. Cambridge University Press (1989)
22. Bak P.: How Nature Works: The Science of Self-Organized Criticality. Springer, New York (1996)
23. Balassa B.: The Purchasing Power Parity Doctrine: A Reappraisal. J. Political Econ. **72**, 584-596 (1964)
24. Baldwin R.: Re-Interpreting the Failure of Foreign Exchange Market Efficiency Tests: Small Transaction Costs, Big Hysterisis Band. Working Paper Series **3319**, National Bureau of Economic Research (1990), http://www.nber.org. Cited 07 Mar 2005
25. Balke N., Wohar M.: Nonlinear Dynamics and Covered Interest Parity. Empir. Econ. **23**, 535-559 (1998)
26. Banerjee A., Dolado J.J., Hendry D.F., Smith G.W.: Exploring Equilibrium Relationships in Econometrics through Static Models: Some Monte Carlo Evidence. Oxford Bull. Econ. Stat. **3**, 253-278 (1986)
27. Banks J., Carson II J.S., Nelson B.L., Nicol D.M.: Discrete Event System Simulation. Third Edition, Prentice-Hall, Upper Saddle River NJ (2000)
28. Battiti R.: Reactive Search: Toward Self-Tuning Heuristics. In: Rayward-Smith V. J., Osman I. H., Reeves C. R., Smith G. D. (eds.) Modern Heuristic Search Methods, pp. 61-83. John Wiley and Sons, Chichester (1996)
29. Baz J., Breedon F., Naik V., Peress J.: Optimal Currency Portfolios: Trading on the Forward Bias. Lehman Brothers Analytical Research Series, October (1999)
30. Bekaert G.: Exchange Rate Volatility and Deviations from Unbiasedness in a Cash-in-Advance Model. J. Int. Econ. **36**, 29-52 (1994)
31. Bekaert G., Hodrick R.J.: Characterizing Predictable Components in Excess Returns on Equity and Foreign Exchange. J. Finance **66**, 251-287 (1992)
32. Bellman R.: Adaptive Control Processes: A Guided Tour. Princeton University Press, New Jersey (1961)
33. Benninga S., Protopapadakis A.A.: The Equilibrium Pricing of Exchange Rates and Assets When Trade Takes Time. J. Int. Econ. **7**, 129-149 (1988)
34. Bertsimas D.: Probabilistic Combinational Optimization Problems. PhD Thesis, Oper. Res. Cent., Massachusetts Institute of Technology, Cambridge MA (1988)
35. Bertsimas D.J., Jaillet P., Odoni A.: A Priori Optimization. Oper. Res. **38**, 1019-1033 (1990)
36. Bianchi L., Dorigo M., Gambardella L.M., Gutjahr W.J.: Metaheuristics in Stochastic Combinatorial Optimization: A Survey. Technical Report IDSIA-08-06, Dalle Molle Institute for Artificial Intelligence, Manno CH (2006), http://www.idsia.ch/idsiareport/IDSIA-08-06.pdf. Cited 11 Oct 2006
37. Bilson J.F.O.: The Speculative Efficiency Hypotheses. J. Bus. **54(3)**, 435-451 (1981)
38. Birge J.R., Louveaux F.: Introduction to Stochastic Programming. Springer, New York NY (1997)
39. Bank for International Settlements: Triennial Central Bank Survey of Foreign Exchange and Derivatives Market Activity in April 2007 - Preliminary global results. Bank for International Settlements, Basel CH (2007), http://www.bis.org/publ/rpfx07.pdf. Cited 27 Jan 2008
40. Black F.: The Pricing of Commodity Contracts. J. Financial Econ. **3**, 167-179 (1976)
41. Black F., Scholes M.: The Pricing of Options and Corporate Liabilities. J. Political Econ. **81(3)**, 637-654 (1973)
42. Blanchard O.J., Quah D.: The Dynamic Effects of Aggregate Demand and Supply Disturbances. Am. Econ. Rev. **79**, 655-673 (1989)
43. Bloomberg L.P.: EUR/USD Forward Exchange Rate Price Data 01/1999 to 12/2006. Bloomberg database. BMW AG, Munich, Germany. 27th August, 2007
44. Bloomberg L.P.: EUR/USD European Style Straddle Option Price Data 01/1999 to 12/2006. Bloomberg database. BMW AG, Munich, Germany. 27th August, 2007
45. Blattberg R.C., Gonedes N.J.: A Comparison of the Stable and Student Distributions as Statistical Models for Stock Prices. J. Bus. **47**, 244-280 (1974)
46. Blum C., Roli A.: Metaheuristics in Combinatorial Optimization: Overview and Conceptual Comparison. ACM Comput. Surv. **35(3)**, 268-308 (2003)

47. Bodnar G.M., Gebhardt G.: Derivatives Usage in Risk Management by US and German Non-Financial Firms: A Comparative Survey. J. Int. Financial Manag. Account. **10**, 153-187 (1999)
48. Bolland P.J., Connor J.T., Refenes A.P.: Application of Neural Networks to Forecast High Frequency Data: Foreign Exchange. In: Dunis C., Zhou B. (eds.) Nonlinear Modelling of High Frequency Financial Time Series, Wiley (1998)
49. Bollerslev T.: Generalised Autoregressive Conditional Heteroskedasticity. J. Econ. **31**, 307-327 (1986)
50. Bollerslev T., Chou R., Kroner K.: ARCH Modelling in Finance. J. Econ. **52**, 5-59 (1986)
51. Bollerslev T., Engle R.F., Nelson D.B.: ARCH Models. In: Engle R.F., McFadden D.L. (eds.) Handbook of Econometrics Vol. 4., Elsevier, North-Holland (1994)
52. Bookstaber R.M., Clarke R.: Options Can Alter Portfolio Return Distributions. J. Portf. Manag. 1981 **7(3)**, 63-70 (1981)
53. Bookstaber R., Clarke R.: Problems in Evaluating the Performance of Portfolios with Options. Financial Analysts J. **41**, 48-62 (1985)
54. Boothe P., Glassman D.: Comparing Exchange Rate Forecasting Models. Int. J. Forecast. **3**, 65-79 (1987)
55. Boser B.E., Guyon I.M., Vapnik V.N.: A Training Algorithm for Optimal Margin Classifiers. In: Haussler D. (ed.) Proceedings of the 5th Annual ACM Workshop on Computational Learning Theory, pp. 144-152. ACM Press (1992)
56. Bousquet O., Herrmann D.: On the Complexity of Learning the Kernel Matrix. Adv. Neural Inf. Proc. Syst. **15**, 399-406 (2002)
57. Bowden R.J.: Forwards or Options? Nesting Procedures for 'Fire and Forget' Commodity Hedging. J. Futures Mark. **14(8)**, 891-913 (1994)
58. Box G.E.P., Jenkins G.M.: Time Series Forecasting: Analysis and Control. Holden-Day, San Francisco CA (1976)
59. Branson W.: The Minimum Covered Interest Differential Needed for International Arbitrage Activity. J. Political Econ. **77(6)**, 1028-1035 (1969)
60. Brock W., Dechert D., Scheinkmann J., LeBaron B.: A Test for Independence Based on the Correlation Dimension. Econ. Rev. **15(3)**, 197-235 (1996)
61. Brock W.A., Hsieh D.A., LeBaron B.: Nonlinear Dynamics, Chaos, and Instability: Statistical theory and Economic Evidence. MIT Press, Cambridge-London (1991)
62. Brooks C.: Introductory Econometrics for Finance. Cambridge University Press (2002)
63. Brown G.W., Khokher Z.: Corporate Risk Management and Speculative Motives. Working Paper, Kenan-Flagler Business School, University of North Carolina at Chapel Hill (2005), http://ssrn.com/abstract=302414. Cited 05 Feb 2006
64. Brown M., Grundy W., Lin D., Cristianini N., Sugnet C., Furey T., Ares M., Haussler D.: Knowledge-Based Analysis of Microarray Gene Expression Data using Support Vector Machines. Proc. Nat. Academy Sci. **97(1)**, 262-267 (2000)
65. Brown G.W., Crabb P.R., Haushalter D.: Are Firms Successful at Selective Hedging? J. Bus., **79(6)**, 2925-2950 (2006)
66. Burstein A., Eichenbaum M., Rebelo S.: Large Devaluations and the Real Exchange Rate. J. Political Econ. **113**, 742-784 (2004)
67. Cai M.-C., Deng X.: Arbitrage in Frictional Foreign Exchange Market. Electron. Notes Theor. Comput. Sci. **78**, 293-302 (2003)
68. Canuto et al. (1999) Canuto S.A., Resende M.G.C., Ribeiro C.C.: Local Search with Perturbations for the Prize-Collecting Steiner Tree Problem in Graphs. Technical Report **99.14.2**, AT&T Labs Research (1999) http://www.research.att.com/mgcr/doc/pcstpls.pdf. Cited 01 Sep 2005
69. Cassel G.: Abnormal Deviations in International Exchanges. Econ. J. **28**, 413-415 (1918)
70. Chang C.-C., Lin C.-J.: LIBSVM: A Library for Support Vector Machines. Technical Report, Department of Computer Sciences and Information Engineering, National Taiwan University, Taipei (2001), Software available at http://www.csie.ntu.edu.tw/c̃jlin/libsvm

71. Chen S.-S., Engel C.: Does 'Aggregation Bias' Explain the PPP Puzzle? Working Paper Series **10304**, National Bureau of Economic Research (2004), http://www.nber.org. Cited 10 May 2005
72. Cheung Y.-W., Lai K.S.: A Fractional Cointegration Analysis of Purchasing Power Parity. J. Bus. Econ. Stat. **11**, 103-112 (1993)
73. Cheung Y.-W., Lai K.S.: Long-Run Purchasing Power Parity during the Recent Float. J. Int. Econ. **34**, 181-192 (1993)
74. Chinn M.D., Meredith G.: Monetary Policy and Long Horizon Uncovered Interest Parity. IMF Staff Papers **51(3)**, 409-430 (2004)
75. Chipman J.S.: The Ordering of Portfolios in Terms of Mean and Variance. Rev. Econ. Stud. **40**, 167-190 (1973)
76. Chopra V., Ziemba W.: The Effect of Error in Means, Variances, and Covariances on Optimal Portfolio Choice. J. Portf. Manag., **Winter: 6-11** (1993)
77. Clare A., Smith P., Thomas S.: UK Stock Returns and Robust Tests of Mean Variance Efficiency. J. Bank. Finance **21**, 641-660 (1997)
78. Clarke R., Kritzman M.: Currency Management: Concepts and Practices. The Research Foundation of the Institute of Chartered Financial Analysts (1996)
79. Clinton K.: Transaction Costs and Covered Interest Arbitrage: Theory and Evidence. J. Polit. Econ., **April**, 350-358 (1988)
80. Clostermann, J., Schnatz B.: The Determinants of the Euro-Dollar Exchange Rate: Synthetic Fundamentals and a Non-Existing Currency. Appl. Econ. Q. **46(3)**, 274-302 (2000)
81. Consiglio A., Zenios S.A.: Designing Portfolios of Financial Products via Integrated Simulation and Optimization Models. Oper. Res. **47(2)**, 195-208 (1999)
82. Corbae D., Ouliaris S.: Cointegration and Tests of Purchasing Power Parity. Rev. Econ. Stat. **70**, 508-511 (1988)
83. Cortes C., Vapnik V.: Support Vector Networks. Mach. Learn. **20**, 273-297 (1995)
84. Crammer K., Keshet J., Singer Y.: Kernel Design Using Boosting. Adv. Neural Inf. Process. Syst. **15**, 537-544 (2002)
85. Cristianini N., Shawe-Taylor J.: Support Vector Machines and Other Kernel-Based Learning Methods. Cambridge University Press (2000)
86. Cristianini N., Shawe-Taylor J., Elisseeff A., Kandola J.: On Kernel-Target Alignment. In: Dietterich T.G., Becker S., Ghahramani Z. (eds.) Advances in Neural Information Processing Systems Vol.14, pp. 367-373. MIT Press, Cambridge MA (2002)
87. Cristianini et al. (2003) Cristianini N., Kandola J., Elisseeff A., Shawe-Taylor J.: On Optimizing Kernel Alignment. In: Advances in Neural Information Processing Systems **13**, pp. 367-373, MIT Press (2003)
88. Cushman D.O., Zha T.: Identifying Monetary Policy in a Small Open Economy under Flexible Exchange Rates. J. Monet. Econ. **39**, 433-448 (1997)
89. Dacco, R. and Satchell, S.: Why do Regime-Switching Models Forecast so Badly? J. Forecast. **18**, 1-16 (1999)
90. Davis G.K.: Income and Substitution Effects for Mean-Preserving Spreads. Int. Econ. Rev. **61(1)**, 131-136 (1989)
91. De Groot C., Wurtz D.: Analysis of Univariate Time Series with Connectionist Nets: A Case Study of Two Classical Examples. Neurocomput. **3**, 177-192 (1991)
92. DeCoste D., Schlkopf B.: Training Invariant Support Vector Machines. Mach. Learn. **46**, 161-190 (2002)
93. DeMarzo P.M., Duffie D.: Corporate Financial Hedging with Proprietary Information. J. Econ. Theory **53**, 261-286 (1991)
94. DeMarzo P.M., Duffie D.: Corporate Incentives for Hedging and Hedge Accounting. Rev. Financial St. **8**, 743-771 (1995)
95. Deng X.T., Li Z.F., Wang S.Y.: On Computation of Arbitrage for Markets with Friction. In: Du D.Z. et al. (eds.) Lecture Notes in Computer Science Vol. 1858 (Proceedings of the 6th Annual International Conference on Computing and Combinatorics), pp. 310-319. Springer, Berlin and Heidelberg (2000)

96. Deng X.T., Li Z.F., Wang S.Y.: Computational Complexity of Arbi-trage in Frictional Security Market. Int. J. Found. Comput. Sci. **3**, 681-684 (2002)

97. Dewdney A.K.: The Turing Omnibus. Computer Science Press (1989)

98. Dickey D.A., Fuller W.A.: Distribution of the Estimators for Autoregressive Time Series with a Unit Root. J. Am. Stat. Assoc. **74**, 427-431 (1979)

99. Diebold F.X.: Empirical Modeling of Exchange Rate Dynamics. Springer, New York (1988)

100. Dornbusch R.: Expectations and Exchange Rate Dynamics. J. Polit. Econ. **84(6)**, 1161-1176 (1976)

101. Dornbusch R., Fischer S.: Exchange Rates and the Current Account. Am. Econ. Rev. **70(5)**, 960-971 (1980)

102. Dornbusch R., Krugman P.: Flexible Exchange Rates in the Short Run. Brookings Papers on Economic Activity **3**, 537-576 (1976)

103. Dowe D.L., Korb K.B.: Conceptual Difficulties with the Efficient Market Hypothesis. In: Dowe D.L. et al. (eds.) Proceedings: Information, Statistics and Induction in Science (ISIS), pp. 200-211. World Scientific, Singapore (1996)

104. Duan K., Keerthi S.S., Poo A.N.: Evaluation of Simple Performance Measures for Tuning SVM Hyperparameters. Neurocomput. **51**, 41-59 (2003)

105. Dufey G., Srinivasulu S.: The Case of Corporate Management of Foreign Exchange Risk. Financial Manag. **12**, 54-62 (1983)

106. Dumas B.: Dynamic Equilibrium and the Real Exchange Rate in Spatially Separated World. Rev. Financial Stud. **5**, 153-180 (1992)

107. Dunis C.L., Williams M.: Modelling and Trading the EUR/USD Exchange Rate: Do Neural Network Models Perform Better? Derivatives Use **8(3)**, 211-239 (2002)

108. Durbin J., Watson G.S.: Testing for Serial Correlation in Least Squares Regression. Biometrika **38**, 159-171 (1951)

109. Eilenberger G., Schulze H., Wittgen R.: Whrungsrisiken, Whrungsmanagement und Devisenkurssicherung von Unternehmungen. Knapp Fritz, Frankfurt (2004)

110. Eisenhauer J.G.: Estimating Prudence. East. Econ. J. **26(4)**, 379-392 (2000)

111. Eisenhauer J.G., Ventura L.: Survey Measures of Risk Aversion and Prudence. Appl. Econ. **35(13)**, 1477-1484 (2003)

112. Elizondo D.: The Linear Separability Problem: Some Testing Methods. IEEE Trans. Neural Netw. **17(2)**, 330-344 (2006)

113. Engel C.: On the Foreign Exchange Risk Premium in a General Equilibrium Model. J. Int. Econ. **32**, 305-319 (1992)

114. Engel C.: The Forward Discount Anomaly and the Risk Premium: A Survey of Recent Evidence. J. Empir. Finance **3(2)**, 123-192 (1996)

115. Engle R.F.: Autoregressive Conditional Heteroscedasticity with Estimates of the Variance of UK Inflation. Econometrica **50**, 987-1008 (1982)

116. Engle R.F., Granger C.W.J.: Co-Integration and Error Correction: Representation, Estimation, and Testing. Econometrica **55**, 251-276 (1987)

117. Engle R.F., Lee G.G.J.: Long Run Volatility Forecasting for Individual Stocks in a One Factor Model. Economics Working Paper Series **93-30**, University of California, San Diego (1993) ftp://weber.ucsd.edu/pub/econlib/dpapers/ucsd9330.ps.gz. Cited 14 Dec 2006

118. Engle R.F., Lee G.G.J.: A Permanent and Transitory Component Model of Stock Return Volatility. In Engle R.F., White H. (eds.) Cointegration, Causality, and Forecasting: A Festschrift in Honour of Clive W.J. Granger, Chapter 20, pp. 475-497. Oxford University Press, Oxford (1999)

119. Engle R.F., Mezrich J.: Grappling with GARCH. RISK, **8(9)**, 112-117 (1995)

120. Engle R.F., Ng V.K.: Measuring and Testing the Impact of News on Volatility. J. Finance **48**, 1749-1778 (1993)

121. Eun C., Resnick B.: International Equity Investment with Selective Hedging Strategies. J. Int. Financial Mark., Inst. Money **7**, 21-42 (1997)

122. Eun C.S., Sabherwal S.: Forecasting Exchange Rates: Do Banks Know Better? Glob. Finance J. **13**, 195-215 (2002)

123. Evans J.R., Laguna M.: OptQuest User Manual. Decisioneering, Denver CO (2004)
124. Fama E.F.: Efficient Capital Markets: A Review of Theory and Empirical Work. J. Finance **25**, 383-414 (1970)
125. Fama E.F.: Forward and Spot Exchange Rates. J. Monet. Econ. **14** November, 319-338 (1984)
126. Fama E.F.: Efficient Capital Markets II. J. Finance **6(5)**, 1575-1617 (1991)
127. Fawcett T.: ROC Graphs: Notes and Practical Considerations for Researchers. Technical Report HPL-2003-4, HP Laboratories, Palo Alto CA (2004)
128. Feenstra R.C., Kendall J.D.: Pass-Through of Exchange Rates and Purchasing Power Parity. J. Int. Econ. **43**, 237-261 (1997)
129. Feldstein M.: The Effects of Taxation on Risk Taking. J. Polit. Econ. **77**, 755-764 (1969)
130. Fishman M.B., Barr D.S., Loick W.J.: Using Neural Nets in Market Analysis. Tech. Anal. Stocks Commod. **9(4)**, 135-138 (1991)
131. Francois D., Wertz V., Verleysen M.: About the Locality of Kernels in High-Dimensional Spaces. In: Proceedings of the International Symposium on Applied Stochastic Models and Data Analysis (ASDMA'2005), pp. 238-245. Brest (2005)
132. Frankel J.A.: In Search of the Exchange Risk Premium: A Six Currency Test of Mean-Variance Efficiency. J. Int. Money Finance **2**, 255-274 (1982)
133. Frankel J.A.: International Capital Mobility and Crowding-out in the U.S. Economy: Imperfect Integration of Financial Markets or of Goods Markets? In: Hafer R.W. (ed.) How open is the U.S. Economy?, pp. 33-67. Lexington Books, Lexington MA (1986)
134. Frankel J.A.: Zen and the Art of Modern Macroeconomics: The Search for Perfect Nothingness. In: Haraf W., Willett T. (eds.) Monetary Policy for a Volatile Global Economy, pp. 117-123. American Enterprise Institute, Washington D.C. (1990)
135. Frankel J.A., Froot K.A.: Using Survey Data to Test Standard Propositions Regarding Exchange Rate Expectations. Am. Econ. Rev. **77(1)**, 133-153 (1987)
136. Frenkel J.A.: Purchasing Power Parity: Doctrinal Perspective and Evi-dence from the 1920s. J. Int. Econ. **8**, 169-191 (1978)
137. Frenkel J.: Flexible Exchange Rates, Prices, and the Role of 'News'. J. Polit. Econ. **89**, 665-705 (1981)
138. Frenkel J., Levich R.: Covered Interest Rate Arbitrage: Unexploited Profits? J. Polit. Econ. **83**, 325-338 (1975)
139. Friedman M., Savage L.J.: The Utility Analysis of Choices Involving Risk. J. Polit. Econ. **56(4)**, 279-304 (1948)
140. Froot K.A.: Currency Hedging over Long Horizons. Working Paper Series **W4355**, National Bureau of Economic Research (1993), http://www.nber.org. Cited 09 Mar 2005
141. Froot K.A., Frankel J.A.: Forward Discount Basis: Is it an Exchange Risk Premium? Q. J. Econ. **104(1)**, 139-161 (1989)
142. Froot K.A., Rogoff K.: Perspectives on PPP and Long-Run Real Exchange Rates. In: Grossmann G., Rogoff K. (eds.) Handbook of International Economics, pp. 1647-1688. North Holland, Amsterdam (1995)
143. Froot K., Thaler R.: Anomalies: Foreign Exchange. J. Econ. Perspect. **4(2)**, 139-161 (1990)
144. Froot K.A., Scharfstein D.S., Stein J.C.: Risk Management: Coordinating Corporate Investment and Financing Policies. J. Finance **48**, 1629-1658 (1993)
145. Froot K., Scharfstein D.S., Stein J.: A Framework for Risk Man-agement. J. Appl. Corp. Finance **Fall**, 22-32 (1994)
146. Fu M.C.: Optimization via Simulation: A Review. Ann. Oper. Res. **53**, 199-248 (1994)
147. Fu M.C.: Simulation Optimization. In: Peters B.A., Smith J.S., Medeiros D.J., Rohrer M.W. (eds.) Proceedings of the 2001 Winter Simulation Conference, pp. 53-61. IEEE Computer Society, Washington DC (2001)
148. Fu M.: Optimization for Simulation: Theory and Practice. INFORMS J. Comput. **14(3)**, 192-215 (2002)
149. Fu M.C., Andradttir S., Carson J.S., Glover F., Harrell C.R., Ho Y.C., Kelly J.P., Robinson S.M.: Integrating Optimization and Simulation: Research and Practice. In: Joines J.A., Barton R.R., Kang K., Fishwick P.A. (eds.) Proceedings of the 32nd Winter Simulation Conference, pp. 610-616. Society for Computer Simulation International, San Diego CA (2000)

150. Fylstra D., Lasdon L., Watson J., Warren A.: Design and Use of the Microsoft Excel Solver. Interfaces **28(5)**, 29-55 (1998)

151. Gallant A.R., Hsieh D., Tauchen G.: On Fitting a Recalcitrant Series: The Pound/Dollar Exchange Rate 1974-1983. In: Barnett W., Powell J., Tauchen G. (eds.) Nonparametric and Semiparametric Methods in Econometrics and Statistics, pp. 199-242. Cambridge University Press (1991)

152. Garey R.M., Johnson D.S.: Computers and Intractability: A Guide to the Theory of NP-Completeness. W.H. Freeman and Company, New York (1979)

153. Geanakoplos J.: Nash and Walras Equilibrium via Brouwer. Econ. Theory **21**, 585-603 (2003)

154. Geczy C., Minton B.A., Schrand C.: Why Firms Use Currency Derivatives. J. Finance **52(4)**, 1323-1354 (1997)

155. Gilbert D.T., Pinel E.C., Wilson T.D., Blumberg S.J., Wheatley T.: Immune Neglect: A Source of Durability Bias in Affective Forecasting. J. Pers. Soc. Psychol. **75**, 617-638 (1998)

156. Gilboa I., Postlewaite A., Schmeidler D.: The Complexity of Utility Maximization. Working Paper, October (2006)

157. Giovannini A., Jorion P.: The Time Variation of Risk and Return in the Foreign Exchange and Stock Markets. J. Finance **44**, 307-326 (1989)

158. Girosi F.: An Equivalence between Sparse Approximation and Support Vector Machines. Neural Comput. **10**, 1455-1480 (1998)

159. Glaum M.: Foreign-Exchange-Risk Management in German Firms Non-Financial Corporations: An Empirical Analysis. In: Frenkel M., Hommel U., Rudolf M. (eds.) Risk Management - Challenge and Opportunity, pp. 537-556. Springer, Berlin and Heidelberg (2000)

160. Glen J.D.: Real Exchange Rates in the Short, Medium, and Long Run. J. Int. Econ. **33**, 147-166 (1992)

161. Glover F.: Heuristics for Integer Programming Using Surrogate Constraints. Decis. Sci. **8(1)**, 156-166 (1977)

162. Glover F.: Future Paths for Integer Programming and Links to Artificial Intelligence. Comput. Oper. Res. **13**, 533-549 (1986)

163. Glover F.: A Template for Scatter Search and Path Relinking. Hao J.K., Lutton E., Ronald E., Schoenauer M., Snyers D. (eds.) Lecture Notes in Computer Science Vol. 1363, pp. 13-54. Springer, Berlin and Heidelberg (1997)

164. Glover F., Laguna M.: Tabu Search. Kluwer Academic Publishers, Dordrecht (1997)

165. Glover et al. (1999) Glover F., Løkketangen A., Woodruff D.: Scatter Search to Generate Diverse MIP Solutions. Research Report, University of Colorado, Boulder CO (1999), http://leeds-faculty.colorado.edu/glover/SS-DiverseMIP.PDF. Cited 16 Feb 2007

166. Glover F., Laguna M., Martí R.: Fundamentals of Scatter Search and Path Relinking. Control and Cybernetics **29(3)**, 653-684 (2000)

167. Glover F., Laguna M., Martí R.: Scatter Search. In: Ghosh A., Tsutsui S. (eds.) Theory and Applications of Evolutionary Computation: Recent Trends, pp. 519-529. Springer (2002)

168. Godel K.: ber formal unentscheidbare Stze der Principia Mathematica und verwandter Systeme I. Monatshefte Math. Phys. **38**, 173-198 (1931)

169. Gourinchas P.-O., Tornell A.: Exchange Rate Puzzles and Distorted Beliefs. J. Int. Econ. **64**, 303-333 (2004)

170. Granger C.W.J.: Investigating Causal Relations by Econometric Models and Cross-Spectral Methods. Econometrica **37**, 424-438 (1969)

171. Granger C.W.J., Anderson A.P.: An Introduction to Bilinear Time Series Models. Vandenhock and Ruprecht, Göttingen (1978)

172. Granger C.W., Newbold P.: Spurious Regressions in Econometrics. J. Econom. **44**, 307-325 (1974)

173. Granger C.W.J., Newbold P.: Spurious Regressions in Econometrics. J. Econ. **2**, 111-120 (1979)

174. Granger C.W. J., Tersvirta T.: Modelling Nonlinear Economic Relationships. Oxford University Press, Oxford UK (1993)

175. Grassberger P., Procaccia I.: Measuring the Strangeness of Strange Attractors. Physica D **9**, 189-208 (1983)

176. Gunn S.R., Brown M., Bossley K.M.: Network Performance Assessment for Neurofuzzy Data Modelling. In: Liu X., Cohen P., Berthold M. (eds.) Lecture Notes in Computer Science Vol. 1280, pp. 313-323. Springer, Berlin and Heidelberg (1997)

177. Hager P.: Corporate Risk Management - Cash Flow at Risk und Value at Risk. Bankakademie-Verlag, Frankfurt am Main (2004)

178. Hakkio C.S.: A Re-Examination of Purchasing Power Parity: A Multi-Country and Multi-Period Study. J. Int. Econ. **17**, 265-277 (1984)

179. Halliwell L.J.: ROE, Utility amd the Pricing of Risk. CAS 1999 Spring Forum, Reinsurance Call Paper Program, 71-135 (1999)

180. Hamilton J.D.: Time Series Analysis. Princeton University Press, Princeton NJ (1994)

181. Haneveld W.K.K., van der Vlerk M.H.: Stochastic Integer Programming: State of the Art. An. Oper. Res. **85**, 39-57 (1999)

182. Hanoch G., Levy H.: The Efficiency Analysis of Choices Involving Risk. Rev. Econ. St. **36**, 335-346 (1969)

183. Hanoch G., Levy H.: Efficient Portfolio Selection with Quadratic and Cubic Utility. J. Bus. **43**, 181-189 (1970)

184. Harrod R.: International Economics. James Nisbet, London (1933)

185. Hausman D.M.: The Inexact and Separate Science of Economics. Cambridge University Press, New York (1992)

186. Hayek F.A.: The Use of Knowledge in Society. Am. Econ. Rev. **35**, 519-530 (1945)

187. Heath D.: A Geometric Framework for Machine Learning. PhD Thesis, John Hopkins University, Baltimore MD (1992)

188. Heath D., Kasif S., Salzberg S.: Induction of Oblique Decision Trees. In: Bajcsy R. (ed.) Proceedings of the Thirteenth International Joint Conference on Artificial Intelligence Vol. 2, pp. 1002-1007. Morgan Kaufmann (1993)

189. Henriksson R., Merton R.: On Market Timing and Investment Performance I. An Equilibrium Theory of Value for Market Forecasts. J. Bus. **54**, 363-406 (1981)

190. Hicks J.R.: Value and Capital. Oxford University Press, London UK (1953)

191. Hill T., o'Connor M., Remus W.: Neural Network Models for Time Series Forecasts. Manag. Sci. **42(7)**, 1082-1092 (1996)

192. Hirsch M.D., Papadimitriou C.H., Vavasis S.A.: Exponential Lower Bounds for Finding Brouwer Fixed Points. J. Complex. **5**, 379-416 (1989)

193. Hodrick R.: The Empirical Evidence on the Efficiency of Forward and Futures Exchange Markets. Harcourt Academic Publishers, Chur CH (1987)

194. Holland J.H.: Complex Adaptive Systems. Daedalus, **121**, 17-20, (1992)

195. Hollifield B., Uppal R.: An Examination of Uncovered Interest Rate Parity in Segmented International Commodity Markets. J. Finance **52**, 2145-2170 (1997)

196. Hornik K., Stinchcombe M., White H.: Multilayer Feedforward Networks are Universal Approximators. Neural Netw. **2**, 359-366 (1989)

197. Hsieh D.A.: Testing for Nonlinear Dependence in Daily Foreign Exchange Rates. J. Bus. **62**, 339-368 (1989)

198. Hsieh D.A.: Chaos and Nonlinear Dynamics: Application to Financial Markets. J. Finance **46(5)**, 1839-1877 (1991)

199. Huizinga J.: An Empirical Investigation of the Long-Run Behaviour of Real Exchange Rates. Carnegie-Rochester Conf. Ser. Pub. Policy **27**, 149-214 (1987)

200. Hull J.C.: Options, Futures, and Other Derivatives. Fourth Edition, Prentice-Hall International, London (2000)

201. Ihaka R., Gentleman R.: A Language for Data Analysis and Graphics. J. Comput. Gr. Stat. **5(3)**, 299-314 (1996)

202. Isukapalli S.S.: Uncertainty Analysis of Transport-Transformation Models. PhD Thesis, Rutgers State University of New Jersey, New Brunswick NJ (1999)

203. Jaakkola T.S., Haussler D.: Exploiting Generative Models in Discriminative Classifiers. In: Kearns M.S., Solla S.A., Cohn D.A. (eds.) Advances in Neural Information Processing Systems, Vol.11. MIT Press, Cambridge MA (1998)

204. Jaillet P.: A Priori Solution of a Traveling Salesman Problem in which a Random Subset of the Customers are Visited. Oper. Res. **36(6)**, 929-936 (1988)
205. Jansen E.S., Tersvirta T.: Testing Parameter Constancy and Super Exogeneity in Econometric Equation. Oxford Bulletin of Economics and Statistics **58**, 735-768 (1996)
206. Jarque C.M., Bera A.K.: Efficient Tests for Normality, Homoscedasticity, and Serial Independence of Regression Residuals. Econ. Lett. **6**, 255-259 (1980)
207. Joachims T.: Making Large-Scale Support Vector Machine Learning Practical. In: Schlkopf B., Burges C., Smola A. (eds.) Advances in Kernel Methods: Support Vector Learning, pp. 169-184. MIT Press, Cambridge MA (1998)
208. Joachims T.: Text Categorization with Support Vector Machines. Proc. Eur. Conf. Mach. Learn. (1998)
209. Johansen S.: Estimation and Hypothesis Testing of Cointegrating Vectors in Gaussian Vector Autoregressive Models. Econometrica **59**, 1551-1580 (1991)
210. Johansen S.: Likelihood-Based Inference in Cointegrated VAR Models. Oxford University Press, Oxford UK (1995)
211. Johnson N.F., Jefferies P., Hui P.M.: Financial Market Complexity: What Physics can Tell us about Market Behaviour. Oxford University Press, Oxford UK (2003)
212. Jorion P.: Value at Risk. Irwin, Chicago (1997)
213. Kahl K.H.: Determination of the Recommended Hedging Ratio. Am. J. Agric. Econ. **650**, 603-605 (1983)
214. Kahneman D., Snell J.: Predicting Utility. In: Hogarth R.M. (ed.) Insights in Decision Making: A Tribute to Hillel J. Einhorn, pp. 295-310. University of Chicago Press, Chicago IL (1990)
215. Kaldor N.: Speculation and Economic Stability. Rev. Econ. St. **7**, 1-27 (1939)
216. Kall P., Wallace S.W.: Stochastic Programming. Wiley, New York NY (1994)
217. Kamruzzaman J., Sarker R.A.: Forecasting of Currency Exchange Rate: A Case Study. In: Proceedings of the IEEE International Conference on Neural Networks and Signal Processing (ICNNSP 2003), pp. 793-797. Nanjing (2003)
218. Kamruzzaman J., Sarker R.A.: Application of Support Vector Machine to Forex Monitoring. IEEJ Trans. Electron. Inf. Syst. **124(10)**, 1944-1951 (2004)
219. Kamruzzaman J., Sarker R.A., Ahmad I.: SVM Based Models for Predicting Foreign Currency Exchange Rates. In: Proceedings of the Third IEEE International Conference on Data Mining (ICDM 2003), pp. 557 (2003)
220. Karatzoglou A., Hornik K., Smola A., Zeileis A.: Kernlab - An S4 Package for Kernel Methods in R. J. Stat. Softw. **11(9)**, 1-20 (2004)
221. Katz J.O.: Developing Neural Network Forecasters for Trading. Techn. Anal. Stocks Commod. **10(4)**, 58-70 (1992)
222. Kaufman L.: Solving the Quadratic Programming Problem Arising in Support Vector Classification. In: Schlkopf B., Burges C., Smola A. (eds.) Advances in Kernel Methods: Support Vector Learning, pp. 147-167. MIT Press (1998)
223. Kean J.: Using Neural Nets for Intermarket Analysis. Techn. Anal. Stocks Commod. **10(11)** (1992)
224. Kennedy M.P., Chua L.O.: Neural Networks for Nonlinear Programming. IEEE Trans. Circuits Syst. **35(3)**, 554-62 (1988)
225. Kenyon A., Morton D.P.: A Survey on Stochastic Loca-tion and Routing Problems. Cent. Eur. J. Oper. Res. **9**, 277-328 (2002)
226. Keynes J.M.: A Treatise on Money, Vol. 2. Macmillan, London UK (1930)
227. Kilian L., Taylor M.P.: Why is it so Difficult to Beat the Random Walk Forecast of Exchange Rates? J. Int. Econ. **60(1)**, 85-107 (2003)
228. Kim Y.: Purchasing Power Parity in the Long Run: A Cointegration Approach. J. Money Credit Bank. **22**, 491-503 (1990)
229. Kimball M.S.: Precautionary Saving in the Small and in the Large. Econometrica **58(1)**, 53-73 (1990)
230. Knetter M.M.: Price Discrimination by U.S. and German Exporters. Am. Econ. Rev. **79**, 198-210 (1989)

231. Koedijk K.G., Tims B., van Dijk M.A.: Purchasing Power Parity and the Euro Area. J. Int. Money Finance **23(7-8)** November-December, 1081-1107 (2004)
232. Korhonen M.: Non-Linearities in Exchange Rate: Evidence from Smooth Transition Regression Model. PhD Thesis, University of Oulu (2005), http://herkules.oulu.fi/isbn9514279468/isbn9514279468.pdf. Cited 29 Jun 2006
233. Kritzman M.P.: The Optimal Currency Hedging Policy with Biased Forward Rates. J. Portf. Manag. **Summer**, 94-100 (1993)
234. Kuan C.M., Liu T.: Forecasting Exchange Rates Using Feed-forward and Recurrent Neural Networks. J. Appl. Econom. **10**, 347-364 (1995)
235. Krugman P., Obstfeld M.: International Economics: Theory and Policy. Third Edition, Harper Collins, New York (1994)
236. Kwiatkowski D., Phillips P.C.B., Schmidt P., Shin Y.: Testing the Null Hypothesis of a Stationary Against the Alternative of a Unit Root. J. Econom. **54**, 159-178 (1992)
237. Ladd G.W., Hanson S.D.: Price-Risk Management with Options: Optimal Market Positions and Institutional Value. J. Futures Mark. **11(6)**, 737-750 (1991)
238. Laguna M., Martí R.: The OptQuest Callable Library. In: Voss S., Woodruff D.L. (eds.) Optimization Software Class Libraries, Operations Research/Computer Science Interfaces Series Vol.18, pp. 193-215. Kluwer Academic Publishers, Boston MA (2002)
239. Laguna M., Martí R.: Scatter Search: Methodology and Implementations in C. Kluwer Academic Publishers, Boston MA (2003)
240. Lai T.: Portfolio Selection with Skewness: A Multiple-Objective Approach. Rev. Quant. Finance Account. **1**, 293-305 (1991)
241. Lanckriet G., Cristianini N., Bartlett P., El Ghaoui L., Jordan M.I.: Learning the Kernel Matrix with Semi-Definite Programming. J. Mach. Learn. Res. **5**, 27-72 (2004)
242. Lapan R., Moschini G., Hanson S.: Production, Hedging, and Speculative Decisions with Options and Futures Markets. Am. J. Agric. Econ. **73**, 745-772 (1991)
243. Laporte G., Louveaux F.V., Mercure H.: A Priori Optimization of the Probabilistic Traveling Salesman Problem. Oper. Res. **42**, 543-549 (1994)
244. Lastrapes W.D.: Sources of Fluctuations in Real and Nominal Exchange Rates. Rev. Econ. Stat. **74**, 530-539 (1992)
245. Latham M.: Defining Capital Market Efficiency. Finance Working Paper **150**, Institute for Business and Economic Research, University of California at Berkeley, April (1985)
246. Lavrac N., Motoda H., Fawcett T., Holte R., Langley P., Adriaans P.: Introduction: Lessons Learned from Data Mining Applications and Collaborative Problem Solving. Mach. Learn. **57(1-2)**, 13-34 (2004)
247. LeCun Y., Jackel L.D., Bottou L., Brunot A., Cortes C., Denker J.S., Drucker H., Guyon I., Mller U.A., Sckinger E., Simard P., Vapnik V.: Learning algorithms for classification: A Comparison on Handwritten Digit Recognition. Neural Netw., 261-276 (1995)
248. Lee T.H., White H., Granger C.W.J.: Testing for Neglected Nonlinearity in Time Series Models. J. Econom. **56**, 269-290 (1993)
249. Leitner J., Schmidt R., Bofinger P.: Biases of Professional Exchange Rate Forecasts: Psychological Explanations and Experimentally Based Comparison to Novices. CEPR Discussion Paper **4230** (2004), http://www.cepr.org/pubs/dps/DP4230.asp. Cited 7 Jul 2005
250. Leland H.E.: Beyond Mean-Variance: Performance Measurement in a Nonsymmetrical World. Financial Analyst J. **55**, 27-35 (1999)
251. Lence S.H., Sakong Y., Hayes D.J.: Multiperiod Production with Forward and Option Markets. Am. J. Agric. Econ. **76**, 286-295 (1994)
252. Lessard D.R., Nohria N.: Rediscovering Functions in the MNC: The Role of Expertise in the Firm's Responses to Shifting Exchange Rates. In: Bartlett C.A., Doz Y., Hedlund G. (eds.) Managing the Global Firm, pp. 186-212. Routledge, London and New York (1990)
253. Leung M.T., Daouk H., Chen A.S.: Using Investment Portfolio Return to Combine Forecasts: A Multiobjective Approach. Eur. J. Oper. Res. **134**, 84-102 (2001)
254. Levi M.: Measurement Errors and Bounded OLS Estimates. J. Econom. **6**, 165-171 (1977)
255. Levich R.M.: Evaluating the Performance of the Forecasters. In: Ensor R. (ed.) The Management of Foreign Exchange Risk, pp. 121-134. Euromoney Publications, London (1982)

256. Levich R., Thomas L.: The Merits of Active Currency Management: Evidence from International Bond Portfolios. Financial Analysts J. **49(5)**, 63-70, (1993)
257. Levy H., Markowitz H.M.: Approximating Expected Utility by a Function of Mean and Variance. Am. Econ. Rev. **69**, 308-317 (1979)
258. Levy H., Sarnat M.: Portfolio and Investment Selection: Theory and Practice. Prentice-Hall, Englewood Cliffs NJ (1984)
259. Lewis A.A.: On Effectively Computable Choice Functions. Math. Soc. Sci. **10**, 43-80 (1985)
260. Lewis A.A.: On Turing Degrees of Walrasian Models and a General Impossibility Result in the Theory of Decision Making. Math. Soc. Sci. **24(2-3)**, 141-171 (1992)
261. Lewis K.K.: Can Learning Affect Exchange-Rate Behaviour? J. Monet. Econ. **23**, 79-100 (1989)
262. Lewis K.K.: Changing Beliefs and Systematic Rational Forecast Errors with Evidence from Foreign Exchange. Am. Econ. Rev. **79(4)**, 621-636 (1989)
263. Lewis K.K.: Puzzles in International Financial Markets. In: Grossman G., Rogoff K. (eds.) Handbook of International Economics Vol. 3, pp. 1913-1971. Elsevier, North-Holland and Amsterdam (1995)
264. Lin C.F., Tersvirta T.: Testing the Constancy of Regression Parameters against Continuous Structural Change. J. Econom. **6**, 211-228 (1994)
265. Liu S.C., Wang S.Y., Qiu W.H.: A Mean-Variance-Skewness Model for Portfolio Selection with Transaction Costs. Int. J. Syst. Sci. **34(4)**, 255-262 (2003)
266. Ljung G., Box G.: On a Measure of Lack of Fit in Time Series Models. Biometrika **66**, 265-270 (1979)
267. Lopez C., Papell D.: Convergence to Purchasing Power Parity at the Commencement of the Euro. Rev. Int. Econ. **15(1)**, 1-16 (2007)
268. Lothian J.R., Taylor M.P.: Real Exchange Rate Behaviour: The Recent Float from the Perspective of the Past Two Centuries. J. Polit. Econ. **104(3)**, 488-509 (1996)
269. Lucas R.E.: Interest Rates and Currency Prices in a Two-Country World. J. Monet. Econ. **10**, 335-359 (1982)
270. Lyons R.K.: The Microstructure Approach to Exchange Rates. MIT Press, Cambridge and London (2001)
271. MacDonald R.: Concepts to Calculate Equilibrium Exchange Rates: An Overview. Economic Research Group of the Deutsche Bundesbank, Discussion Paper **3/00** (2000), http://217.110.182.54/download/volkswirtschaft/dkp/2000/200003dkp.pdf. Cited 17 Aug 2004
272. MacKinlay A.C., Richarson M.P.: Using Generalized Method of Moments to Test Mean-Variance Efficiency. J. Finance **46(2)**, 511-527 (1991)
273. Mandelbrot B.: The Variation of Certain Speculative Prices. J. Bus. **36**, 394-419 (1963)
274. Mangasarian O.L., Musicant D.R.: Successive Overrelaxation for Support Vector Machines. IEEE Trans. Neural Netw. **10(5)**, 1032-1037 (1999)
275. Mark N.: Some Evidence on the International Inequality of Real Interest Rates. J. Int. Money Finance **4**, 189-208 (1985)
276. Mark N.: Real and Nominal Exchange Rates in the Long Run: An Empirical Investigation. J. Int. Econ. **28(1-2)**, 115-136 (1990)
277. Mark N.C., Moh Y.-K.: Continuous-Time Market Dynamics, ARCH Effects, and the Forward Premium Anomaly. Working Paper, University of Notre Dame and Tulane University (2002)
278. Markowitz H.: Portfolio Selection. J. Finance **7**, 77-91 (1952)
279. Markowitz H.: Portfolio Selection: Efficient Diversification of Investments. John Wiley and Sons, New York NY (1959)
280. Markowitz H.M.: Mean-Variance Analysis in Portfolio Choice and Capital Markets. Basil Blackwell, Oxford UK (1987)
281. Markowitz H.: Portfolio Selection: Efficient Diversification of Investments. Second Edition, Blackwell, Cambridge MA (1991)
282. Maurer R., Valiani S.: Hedging the Exchange Rate Risk in International Portfolio Diversification: Currency Forwards versus Currency Options. Working Paper, Goethe University, Frankfurt am Main (2003)

283. Mavrides N.: Triangular Arbitrage in a Foreign Exchange Market - Inefficiencies, Technology and Investment Opportunities. Quorum Books, London UK (1992)
284. Mayers D., Smith C.W.: On the Corporate Demand for Insurance. J. Bus. **55**, 281-296 (1982)
285. McCallum B.T.: Monetary Economics: Theory and Policy. Prentice-Hall, New York and London (1994)
286. McLeod A.I., Li W.K.: Diagnostic Checking ARMA Time Series Models Using Squared-Residual Autocorrelations. J. Time Ser. Anal. **4**, 269-273 (1983)
287. Meese R.A., Rogoff K.: Empirical Exchange Rate Models of the Seventies: Do They Fit Out of Sample? J. Int. Econ. **14**, 793-815 (1983)
288. Megiddo N.: On the Complexity of Polyhedral Separability. Discret. Computat. Geom. **3**, 325-337 (1988)
289. Mercer J.: Functions of Positive and Negative Type and their Connection with the Theory of Integral Equations. Philos. Trans. Royal Soc., **A 209**, 415-446 (1909)
290. Merrigan P., Normandin M.: Precautionary Saving Motives: An Assessment from UK Time Series of Cross-Sections. Econ. J. **106**, 1193-1208 (1996)
291. Meyer D., Leisch F., Hornik K.: The Support Vector Machine under Test. Neurocomput. **55(1-2)**, 169-186 (2003)
292. Michael P., Nobay A.R., Peel D.A.: Transaction Costs and Nonlinear Adjustment in Real Exchange Rates: An Empirical Investigation. J. Polit. Econ. **105(4)**, 862-879 (1997)
293. Michaud R.O.: The Markowitz Optimization Enigma: Is 'Optimized' Optimal? Financial Analysts J. **45**, 31-42 (1989)
294. Mina J., and Xiao J.Y.: Return to RiskMetrics: The Evolution of a Standard. RiskMetrics Group, New York (2001) http://www.riskmetrics.com. Cited 13 Nov 2004
295. Minka T.P.: Empirical Risk Minimization is an Incomplete Inductive Principle. 2001, http://www.citeseer.ist.psu.edu/345924.html. Cited 09 May 2005
296. Minoux M.: Mathematical Programming: Theory and Algorithms. John Wiley and Sons, Chichester (1986)
297. Mitchell T.M.: Machine Learning, McGraw-Hill (1997)
298. Modigliani F., Miller M.: The Cost of Capital, Corporation Finance and the Theory of Investment. Am. Econ. Rev. **48** June, 261-97 (1958)
299. Moriarty E., Phillips S., Tosini P.: A Comparison of Options and Futures in the Management of Portfolio Risk. Financial Analysts J. **37**, 61-67 (1981)
300. Moscato P.: Memetic Algorithms: A Short Introduction. In: Corne D., Dorigo M., Glover F. (eds.) New Ideas in Optimization, pp. 219-234. McGraw-Hill, London UK (1999)
301. Müller K.-R., Smola A., Rätsch G., Schölkopf B., Kohlmorgen J., Vapnik V.: Using Support Vector Machines for Time Series Prediction. In: Sch"lkopf B., Burges C.J.C., Smola A.J. (eds.) Advances in Kernel Methods, pp. 242-253. MIT Press, Cambrdige MA (1999)
302. Murray C., Papell D.: The Purchasing Power Parity is Worse Than You Think. Empir. Econ. **30(3)**, 783-790 (2005)
303. Murthy S.K., Kasif S., Salzberg S.: A System for Induction of Oblique Decision Trees. J. Artif. Intell. Res. **2**, 1-32 (1994)
304. Muth J.: Rational Expectations and the Theory of Price Movements. Econometrica **29**, 315-335 (1961)
305. Myers R.J., Thompson S.R.: Generalized Optimal Hedge Ratio Estimation. Am. J. Agric. Econ. **71**, 858-868 (1989)
306. Nabney I., Dunis C., Rallaway R., Leong S., Redshaw W.: Leading Edge Forecasting Techniques for Exchange Rate Prediction. Eur. J. Finance **1(4)**, 311-323 (1995)
307. Nance D.R., Smith C.W., Smithson C.W.: On the Determinants of Corporate Hedging. J. Finance **48**, 267-284 (1993)
308. Nash Jr. J.F.: Equilibrium Points in n-Person Games. Proceedings of the National Academy of Sciences U.S.A. **36**, 48-49 (1950)
309. Nautz D., Offermanns C.J.: Does the Euro Follow the German Mark? Evidence from the Monetary Model of the Exchange Rate. Eur. Econ. Rev. **50(5)**, 1279-1295 (2006)
310. Nelson C.R.: The Prediction Performance of the F.R.B.-M.I.T.-PENN Model of the U.S. Economy. Am. Econ. Rev. **62**, 902-917 (1972)

311. Obstfeld M., Rogoff K.: Exchange Rate Dynamics Redux. J. Polit. Econ. **103**, 624-660 (1995)
312. Obstfeld M., Rogoff K.: The Six Major Puzzles in International Macroeconomics: Is there a Common Cause? Working Paper Series **7777**, National Bureau of Economic Research (2000), http://www.nber.org. Cited 03 Apr 2006
313. Obstfeld M., Taylor A.M.: Nonlinear Aspects of Goods-Market Arbitrage and Adjustment: Heckscher's Commodity Points Revisited. J. Jpn. Int. Econ. **11**, 441-479 (1997)
314. Ong C.S., Smola A.J., Williamson R.C.: Learning the Kernel with Hyperkernels. J. Mach. Learn. Res. **6**, 1043-1071 (2005)
315. Ortobelli S., Rachev S.T., Stoyanov S., Fabozzi F.J., Biglova A.: The Proper Use of Risk Measures in Portfolio Theory. Int J. Theoret. Appl. Finance **8**, 1-27 (2005)
316. Osman I.H., Laporte G.: Metaheuristic: A Bibliogaphy. An. Oper. Res. **63**, 513-623 (1996)
317. Osuna E., Freund R., Girosi F.: An improved Training Algorithm for Support Vector Machines. In: Principe J., Gile L., Morgan N., Wilson E. (eds.) Neural Networks for Signal Processing Vol. 7 (Proceedings of the 1997 IEEE Workshop), pp. 276-285, New York NY (1997)
318. Palm F.C.: GARCH Models of Volatility. In: Maddala G.S., Rao C.R. (eds.) Handbook of Statistics Vol.14, pp. 209-240. Elsevier, North-Holland and Amsterdam (1996)
319. Papadimitriou C.H.: On the Comlexity of the Parity Argument and Other Inefficient Proofs of Existence. J. Comput. Syst. Sci. **48**, 498-532 (1994)
320. Papadimitriou C.H.: Computational Complexity. Addison Wesley (1994)
321. Peck A.E.: Hedging and Income Stability: Concepts, Implications, and an Example. Am. J. Agric. Econ. **57**, 410-419 (1975)
322. Perold A.F., Schulman E.: The Free Lunch in Currency Hedging: Implications for Investment Policy and Performance Standards. Financial Anal. J. **44(3)**, 45-50 (1988)
323. Pfennig M.: Optimale Steuerung des Whrungsrisikos mit derivativen Instrumenten. Beitrge zur betriebswirtschaftlichen Forschung **83**, Gabler, Wiesbaden (1998)
324. Phillips P.C.B., Perron P.: Testing for a Unit Root in Time Series Regression. Biometrika **75**, 335-346 (1988)
325. Pippenger J.: Arbitrage and Efficient Markets Interpretation of Purchasing Power Parity: Theory and Evidence. Econ. Rev., Federal Re-serve Bank of San Francisco, 31-47 (1986)
326. Platt J.C.: Fast Training of Support Vector Machines Using Sequential Minimal Optimization. In: Schlkopf B., Burges C., Smola A. (eds.) Advances in Kernel Methods, pp. 185-208. MIT Press, Cambridge MA (1998)
327. Plutowski M., White H.: Selecting Concise Training Sets from Clean Data. IEEE Trans. Neural Netw. **4(2)**, 305-318 (1993)
328. Pontil M., Verri A.: Object Recognition with Support Vector Machines. IEEE Trans. PAMI **20**, 637-646 (1998)
329. Post E.: Recursively Enumerable Sets of Positive Integers and Their Decision Problems. Bull. Am. Math. Soc. **50**, 284-316 (1944)
330. Pratt J.W.: Risk Aversion in the Small and in the Large. Econometrica **32(1-2)**, 122-136 (1964)
331. Pringle J., Connolly R.A.: The Nature and Causes of Foreign Currency Exposure. J. Appl. Corp. Finance **6(3)**, 61-72 (1993)
332. Pringle J.J., Connolly R.A.: The Nature and Causes of Foreign Currency Exposure. In: Brown G.W., Chew D.H. (eds.) Corporate Risk - Strategies and Management, pp. 141-152. Risk Books, London UK (1999)
333. Quinlan M.J., Chalup S.K., Middleton R.H.: Application of SVMs for Colour Classification and Collision Detection with AIBO Robots. Adv. Neural Inf. Process. Syst. **16**, 635-642 (2004)
334. Rachev S.T., Ortobelli S., Stoyanov S., Fabozzi F.: Desirable Properties of an Ideal Risk Measure in Portfolio Theory. Working Paper, University of California, Santa Barbara, University of Karlsruhe, Germany, University of Bergamo IT, FinAnalytica USA, Yale University USA (2005), http://www.pstat.ucsb.edu/research/papers/Desproplv.pdf. Cited 10 Oct. 2007

335. Ramsey J.B.: Tests for Specification Errors in Classical Linear Least Squares Regression Analysis. J. Royal Stat. Soc., Series B, **31**, 350-371 (1969)
336. Refenes A.-P., Abu-Mostafa Y., Moody J.: Neural Networks in Financial Engineering. In: Weigend A. (ed.) Proceedings of the Third International Conference on Neural Networks in the Capital Markets, World Scientific, River Edge NJ (1996)
337. Rego C.: Integrating Advanced Principles of Tabu Search for the Vehicle Routing Problem. Working Paper, Faculty of Sciences, University of Lisbon (1999)
338. Rochat Y., Taillard E.D.: Probabilistic Diversification and Intensification in Local Search for Vehicle Routing. J. Heuristics **1**, 147-167 (1995)
339. Rockafellar R.T., Uryasev S.: Optimization of Conditional Value-at-Risk. J. Risk **2**, 21-41 (2000)
340. Rogers J.H.: Monetary Shocks and Real Exchange Rates. J. Int. Econ., **49**, 269-288 (1999)
341. Rogoff K.: Expectations and Exchange Rate Volatility. PhD Thesis, Massachusetts Institute of Technology (1979)
342. Rogoff K.: The Purchasing Power Parity Puzzle. J. Econ. Lit. **34(2)**, 647-668 (1996)
343. Roll R.: Violations of Purchasing Power Parity and their Implications for Efficient International Commodity Markets. In: Sarnat M., Szego G.P. (eds.) International Finance and Trade Vol. 1, 133-176. Ballinger, Cambridge MA (1979)
344. Rosenberg M.R., Folkerts-Landau D.: The Deutsche Bank Guide to Exchange-Rate Determination: A Survey of Exchange-Rate Forecasting Models and Strategies. Deutsche Bank Global Markets Research, New York (2002)
345. Rosenblatt, A.: Principles of Neurodynamics. Spartan, New York NY (1959)
346. Roth A.: The Economist as Engineer: Game Theory, Experimentation, and Computation as Tools for Design Economics. Econometrica **70**, 1341-1378 (2002)
347. Rothschild M., Stiglitz J.: Increasing Risk I: A Definition. J. Econ. Theory **2**, 225-243 (1970)
348. Rothschild M., Stiglitz J.E.: Increasing Risk II: Its Economic Consequences. J. Econ. Theory **3**, 225-243 (1971)
349. Rubinstein M.: Securities Market Efficiency in an Arrow-Debreu Economy. Am. Econ. Rev. **December**, 812-824 (1975)
350. Rubinstein M., Leland H.: Replicating Options with Positions in Stock and Cash. Financial Analysts J. **37**, 63-75 (1981)
351. Rust J.: Dealing with the Complexity of Economic Calculations. In: Durlauf S., Traub J. (eds.) Limits to Knowledge in Economics. Addison-Wesley, Boston MA (1998)
352. Sakong Y., Hayes D.J., Hallam A.: Hedging Production Risk with Options. Am. J. Agric. Econ. **75**, 408-415 (1993)
353. Samuelson P.A.: Theoretical Notes on Trade Problems. Rev. Econ. Stat. **46**, 145-154 (1964)
354. Santa-Clara and Saretto (2005) Santa-Clara P., Saretto A.: Option strategies: Good Deals and Margin Calls. Working Paper, UCLA (2005), http://www.econ.yale.edu/ shiller/behfin/2005-04/santa-clara-saretto.pdf, Cited 03 May 2007
355. Sarantis N.: Modelling Nonlinearities in Effective Exchange Rates. J. Int. Money Finance **18**, 27-45 (1999)
356. Sarno L., Taylor M.P.: Real Exchange Rates under the Recent Float: Unequivocal Evidence of Mean Reversion. Econ. Lett. **60**, 131-137 (1998)
357. Sarno L., Taylor M.P.: The Economics of Exchange Rates. Cambridge University Press, Cambridge UK (2002)
358. Schelling T.C.: Micromotives and Macrobehavior. Norton, New York NY (1978)
359. Schnatz B.: Is reversion to PPP in Euro Exchange Rates Non-Linear? Working Paper **682**, European Central Bank, Frankfurt am Main (2006)
360. Schölkopf B.: The Kernel Trick for Distances. Technical Report MSR 2000-51, Microsoft Research, Redmond WA (2001)
361. Schölkopf B., Smola A.: Learning with Kernels - Support Vector Machines, Regularization, Optimization and Beyond. MIT Press, Cambridge MA (2002)
362. Scott R.C., Horvath P.A.: On the Direction of Preference for Moments of Higher Order than the Variance. J. Finance **35(4)**, 915-919, (1980)

363. Sercu P., Wu X.: Uncovered Interest Arbitrage and Transaction Costs: Errors-in-Variable versus Hysteresis Effects. Working Paper, University of Leuven and City University of Hong Kong (2000)

364. Servaes H., Tufano P.: The Theory and Practice of Corporate Risk Management Policy. Global Markets Liability Strategies Group, Deutsche Bank AG, February (2006)

365. Shih Y.L.: Neural Nets in Technical Analysis. Techn. Anal. Stocks Commod. 9(2), 62-68 (1991)

366. Shiller R.J.: Speculative Prices and Popular Models. J. Econ. Perspect. 4(2), 55-65 (1990)

367. Shiller R.J., Perron P.: Testing the Random Walk Hypothesis: Power Versus Frequency of Observation. Econ. Lett. 18, 381-386 (1985)

368. Sideris D.A.: Foreign Exchange Intervention and Equilibrium Real Exchange Rates. Working Paper 56, Bank of Greece (2007)

369. Simon H.: Models of Man: Social and Rational. Wiley, New York NY (1957)

370. Smith C.W., Stulz R.M.: The Determinants of Firms' Hedging Policies. Financial Quant. Anal. 20, 341-406 (1985)

371. Smola A.J., Schlkopf B.: On a Kernel-Based Method for Pattern Recognition, Regression, Approximation and Operator Inversion. Algorithmica 22, 211-231 (1998)

372. Sortino F.A., Forsey H.J.: On the Use and Misuse of Downside Risk. J. Portf. Manag. 22, 35-42 (1996)

373. Sortino F., Price L.: Performance Measurement in a Downside Risk Framework. J. Invest. 3(3), 59-65 (1994)

374. Spear S.: Learning Rational Expectations Under Computability Constraints. Econometrica 57, 889-910 (1989)

375. Steurer E.: konometrische Methoden und maschinelle Lernverfahren zur Wechselkursprognose. Physica, Wirtschaftswissenschaftliche Beitrge 143, Heidelberg (1997)

376. Stulz R.M.: Rethinking Risk Management. J. Appl. Corp. Finance 9, 8-24 (1996)

377. Swales G.S., Yoon Y.: Applying Artificial Neural Networks to Investment Analysis. Financial Analyst J. 48(5), 70-80 (1992)

378. Sweeney R.J.: Beating the Foreign Exchange Market. J. Finance March, 163-182 (1986)

379. Tang Z., Almedia C., Fishwick P.A.: Time series Forecasting Using Neural Networks vs. Box-Jenkins Methodology. Simul. 57, 303-310 (1991)

380. Taylor M.P.: Covered Interest Parity: A High-Frequency, High-Quality Data Study. Economica 54, 429-438 (1987)

381. Taylor M.P.: Covered Interest Arbitrage and Market Turbulence. Econ. J. 99, 376-391 (1989)

382. Taylor A.M.: A Century of Purchasing Power Parity. Rev. Econ. Stat. 84(1), 139-150 (2002)

383. Taylor M.P.: Is Official Exchange Rate Intervention Effective? Economica 71(1), 1-12 (2004)

384. Taylor M.P., Sarno L.: The Behaviour of Real Exchange Rates During the Post-Bretton Woods Period. J. Int. Econ. 46, 281-312 (1998)

385. Taylor A.M., Taylor M.P.: The Purchasing Power Parity Debate. J. Econ. Perspect. 18(4), 135-58 (2004)

386. Taylor M.P., Peel D.A., Sarno L.: Nonlinear Mean-Reversion in Real Exchange Rates: Toward a Solution to the Purchasing Power Parity Puzzles. Int. Econ. Rev. 42(4), 1015-1042 (2001)

387. Teräsvirta T.: Specification, Estimation, and Evaluation of Smooth Transition Autoregressive Models. J. Am. Stat. Assoc. 89, 208-218 (1994)

388. Teräsvirta T.: Modelling Economic Relationships with Smooth Transition Regressions. In: Ullah A., Giles D.E.A. (eds.) Handbook of Applied Economic Statistics, pp. 507-552. Marcel Dekker, New York (1998)

389. Teräsvirta T., Anderson H.M.: Characterising Nonlinearities in Business Cycles using Smooth Transition Autoregressive Models. J. Appl. Econ. 7, 119-136 (1992)

390. Teräsvirta T., Tjostheim D., Granger C.W.J.: Aspects of Modelling Nonlinear Time Series. In: Engle R.F., McFadden D.L. (eds.) Handbook of Econometrics Vol. 4, pp. 2917-2957. Elsevier Science, Amsterdam (1994)

391. Tobin J.: Liquidity Preference as Behaviour towards Risk. Rev. Econ. St. 25, 65-86 (1958)

392. Tong H.: Threshold Models in Non-Linear Time Series Analysis. Springer, New York NY (1983)

393. Tong H.: Nonlinear Time Series: A Dynamical System Approach. Clarendon Press, Oxford UK (1990)

394. Topaloglou N., Vladimirou H., Zenios S.A.: A Dynamic Stochastic Programming Model for International Portfolio Management. Eur. J. Oper. Res. **185(3)**, 1501-1524 (2008)

395. Topaloglou N., Vladimirou H., Zenios S.A.: Controlling Currency Risk with Options or Forwards. In: Zopounidis C., Doumpos M., Pardalos P.M. (eds.) Handbook of Financial Engineering, Springer (2008)

396. Torniainen A.: Foreign Exchange Risk on Competitive Exposure and Strategic Hedging. PhD Thesis, Helsinki School of Economics and Business Administration (1992)

397. Tsuda K., Akaho S., Asai K.: The EM Algorithm for Kernel Matrix Completion with Auxiliary Data. J. Mach. Learn. Res. **4**, 67-81 (2003)

398. Turing A.M.: On Computable Numbers, with an Application to the Entscheidungsproblem. Proceedings of the London Mathematical Society **42**, 230-265 (1937), and **43**, 544-546 (1937)

399. Ullrich C., Seese D.: Das Beste beider Welten - Bewertungsorientiertes versus Erwartungsorientiertes Whrungsmanagement: Alternative oder Ergnzungsmglichkeit. RISKNEWS **2**, 43-50 (2005)

400. Ullrich C., Seese D., Chalup S.: Predicting Foreign Exchange Rate Return Directions with Support Vector Machines. In: Simoff S.J., Williams G.J., Galloway J., Kolyshkina I. (eds.) Proceedings of the Fourth Australasian Data Mining Conference, pp. 221-240. Sydney Australia (2005)

401. Ullrich C., Seese D., Chalup S.: Foreign Exchange Trading with Support Vector Machines. In: Decker R., Lenz H.-J. (eds.) Advances in Data Analysis: Studies in Classification, Data Analysis, and Knowledge Organization, pp. 539-546. Springer, Berlin and Heidelberg (2006)

402. Ullrich C., Seese D., Chalup S.: Investigating FX Market Efficiency with Support Vector Machines. Paper and Slides submitted to Quantitative Methods in Finance Conference (QMF), Sydney (2006), http://www.business.uts.edu.au/qfrc/qmf/2006//downloads/detlef_seese.pdf. Cited 17 Aug 2007

403. van Dijk D., Tersvirta T., Franses P.H.: Smooth Transition Autoregressive Models - A Survey of Recent Developments. Econ. Rev. **21(1)**, 1-47 (2002)

404. Vapnik V.: The Nature of Statistical Learning Theory. Springer, New York NY (1995)

405. Vapnik V.: Statistical Learning Theory. Wiley (1998)

406. Vapnik V., Golowich S., Smola A.: Support Vector Method for Function Approximation, Regression Estimation, and Signal Processing. In: Mozer M., Jordan M., Petsche T. (eds.) Advances in Neural Information Processing Systems Vol .9, pp. 281-287. MIT Press, Cambridge MA (1997)

407. Velupillai K.: Computable Economics. Arne Ryde Lectures, Oxford University Press, Oxford (2000)

408. Volosov K., Mitra G., Spagnolo F., Lucas C.: Treasury Management Model with Foreign Exchange Exposure. Comput. Optim. Appl. **32(1-2)**, 179-297 (2005)

409. Neumann J.: Zur Theorie der Gesellschaftsspiele. Math. Annalen **100(1)**, 295-320 (1928)

410. Neumann J.: A Model of General Equilibrium. Rev. Econ. St. **13(1)**, 1-9 (1945)

411. Neumann J. von, Morgenstern O.: Theory of Games and Economics Behavior. Princeton University Press, Princeton NY (1947)

412. Wahba G.: Support Vector Machines, Reproducing Kernel Hilbert Spaces and the Randomized GACV. In: Schlkopf B., Burges C.J.C., Smola A.J. (eds.) Advances in Kernel Methods - Support Vector Learning, pp. 69-88. MIT Press, Cambridge MA (1999)

413. Walras L.: Elements of Pure Economics. English translation by William Jaffé, Allen and Unwin (1954)

414. Wang S.Y., Xia Y.S.: Portfolio Selection and Asset Pricing. Springer, Berlin (2002)

415. Weston J., Perez-Cruz F., Bousquet O., Chapelle O., Elisseeff A., Schlkopf B.: Feature Selection and Transduction for Prediction of Molecular Bioactivity for Drug Design. Bioinformatics **19(6)**, 764-771 (2003)

416. White H.: Economic Prediction Using Neural Networks: The Case of IBM Daily Stock Returns. Proceedings of the IEEE International Conference on Neural Networks, pp. 451-458. IEEE Press, New York (1988)

417. Williams C.K.I., Barber D.: Bayesian Classification with Gaussian Processes. IEEE Trans. Pattern Anal. Mach. Intell. PAMI **20(12)**, 1342-1351 (1998)

418. Williams C.K.I., Rasmussen C.E.: Gaussian Processes for Regression. In: Touretzky D.S., Mozer M.C., Hasselmo M.E. (eds.) Advances in Neural Information Processing Systems Vol. 8, pp. 514-520. MIT Press, Cambridge MA (1996)

419. Williams J.C., Wright B.D.: Storage and Commodity Markets. Cambridge University Press, Cambridge UK (1991)

420. Wolpert D.H.: Stacked Generalization. Neural Netw. **5**, 241-259 (1992)

421. Wong F.S.: Fuzzy Neural Systems for Stock Selection. Financial Analyst J. **48**, 47-52 (1992)

422. Working H.: Futures Trading and Hedging. Am. Econ. Rev. **43**, 544-561 (1953)

423. Xu J., Chiu S., Glover F.: Tabu Search and Evolutionary Scatter Search for 'Tree-Star' Network Problems, with Applications to Leased-Line Network Design. In: Corne D.W.W., Oates M.J., Smith G.D. (eds.) Telecommunications Optimization: Heuristic and Adaptive Techniques, pp. 57-79. John Wiley and Sons (2001)

424. Zhang G.: An Investigation of Neural Networks for Linear Time-Series Forecasting. Comput. Oper. Res. **28**, 183-202 (2001)

425. Zhang G.: Time Series Forecasting Using a Hybrid ARIMA and Neural Network Model. Neurocomput. **50**, 159-175 (2003)

426. Zhang G., Patuwo E.B., Hu M.Y.: Forecasting with Artificial Neural Network: The State of the Art. Int. J. Forecast. **14**, 35-62 (1998)